T0331132

Structural Dynamics in Uncertain Environments

The uncertainties or randomness of the material properties of structural components are of serious concern to engineers. Structural analysis is usually done by taking into account only deterministic or crisp parameters; however, building materials can have the aspects of uncertainty. The causes of this uncertainty or randomness are defects in atomic configurations, measurement errors, environmental conditions, and other factors. The influence of uncertainties is more profound for nanoscale and microstructures due to their small-scale effects. Several nanoscale experiments and molecular dynamics studies also support the claim of possible attachment of randomness for various parameters. With regard to these concerns, it is necessary to propose new models that specifically integrate and effectively overcome imprecisely defined parameters of the system.

Structural Dynamics in Uncertain Environments presents the uncertainty modeling of nanobeams, microbeams, and Functionally Graded (FG) beams using non-probabilistic approaches which include interval and fuzzy concepts. Vibration and stability analyses of the beams are conducted using different analytical, semi-analytical, and numerical methods for finding exact and/or approximate solutions of governing equations arising in uncertain environments. In this context, this book addresses structural uncertainties and investigates the dynamic behavior of micro-, nano-, and FG beams.

- Examines the concepts of fuzzy uncertain environments in structural dynamics.
- Presents a comprehensive analysis of propagation of uncertainty in vibration and buckling analyses.
- Explains efficient mathematical methods to handle uncertainties in the governing equations.

Structural Dynamics in Uncertain Environments

Micro, Nano, and Functionally Graded Beam Analysis

Subrat Kumar Jena and S. Chakraverty

CRC Press
Taylor & Francis Group
Boca Raton London New York

CRC Press is an imprint of the
Taylor & Francis Group, an **informa** business

Designed cover image: Shutterstock

First edition published 2025
by CRC Press
2385 NW Executive Center Drive, Suite 320, Boca Raton FL 33431

and by CRC Press
4 Park Square, Milton Park, Abingdon, Oxon, OX14 4RN

CRC Press is an imprint of Taylor & Francis Group, LLC

ISBN: 978-1-032-29494-0 (hbk)
ISBN: 978-1-032-30025-2 (pbk)
ISBN: 978-1-003-30310-7 (ebk)

DOI: 10.1201/9781003303107

Typeset in Times
by Apex CoVantage, LLC

Contents

Author Biography

Dr. Subrat Kumar Jena is presently working as Postdoctoral Fellow in the Department of Applied Mechanics in the Indian Institute of Technology Delhi, New Delhi, India. His academic journey includes an enriching tenure as Honorary Postdoctoral Fellow at the Nonlinear Multifunctional Composites – Analysis & Design (NMCAD) Lab, Department of Aerospace Engineering, Indian Institute of Science (IISc), Bengaluru, India, under the mentorship of Prof. Dineshkumar Harursampath. Dr. Jena holds a Ph.D. from the Department of Mathematics, National Institute of Technology Rourkela, Odisha, India, supervised by Prof. S. Chakraverty. His research expertise encompasses Computational Solid Mechanics, Multifunctional Materials, Applied Mathematics, Mathematical Modeling, and Uncertainty Quantification. Dr. Jena earned his M.Sc. in Mathematics and Scientific Computing from Motilal Nehru National Institute of Technology Allahabad (NIT Allahabad), Prayagraj, Uttar Pradesh, India. His research contributions are widely recognized, with 35 research papers in peer-reviewed international journals, 2 international conference papers, 9 book chapters, and 4 books (2 other books are ongoing) to his credit. Recently, Dr. Jena has been honored with the "Mathematics 2022 Best PhD Thesis Award" by the Mathematics Journal, Basel, Switzerland, for his exceptional research work. This accolade came with a prize of CHF 500 and a certificate, marking a significant milestone in his academic journey. In addition, Dr. Jena has received an IOP Publishing Top-Cited Paper Award in 2021 and 2022 from India, published across the entire IOP Publishing journal portfolio in the past three years, i.e., 2018–2020 and 2019–2021, respectively, for his paper that ranked among the top 1% of the most-cited papers in the materials subject category. In 2020–2021, one of his papers published in ZAMM—*Journal of Applied Mathematics and Mechanic* (Wiley) was among the most-cited papers. Furthermore, three of his papers have been recognized as the best-cited papers published in the *Curved and Layered Structures Journal* (De Gruyter). His contributions to the field of shell buckling, particularly in the area of static or dynamic analysis of general structures, have been acknowledged, as he is included in the Shell Buckling website as "Shell Buckling People". He has also featured in Stanford University's Top 2% Most Influential Scientists List multiple times. Dr. Jena's eminence as a reviewer and guest editor for many prestigious international journals is underscored by his review of over 100 manuscripts.

Prof. S. Chakraverty has 30 years of experience as a researcher and teacher. Presently, he is working in the Department of Mathematics (Applied Mathematics Group), National Institute of Technology, Rourkela, Odisha, as Senior (higher administrative grade) Professor. Prior to this, he was with the CSIR-Central Building Research Institute, Roorkee, India. After completing graduation from St. Columba's College (Ranchi University), his career started from the University of Roorkee (now Indian Institute of Technology Roorkee), and he did M.Sc. (Mathematics) and M.Phil. (Computer Applications) from the said institute securing the first positions in the university. Dr. Chakraverty received his Ph.D. from IIT Roorkee in 1993. Thereafter, he did his postdoctoral research at the Institute of Sound and Vibration Research (ISVR), University of Southampton, UK, and at the Faculty of Engineering and Computer Science, Concordia University, Montreal, Quebec, Canada. He was also Visiting Professor at Concordia and McGill Universities, Canada, during 1997–1999 and Visiting Professor at the University of Johannesburg, Johannesburg, South Africa, during 2011–2014. He has authored/coauthored/edited 31 books, published 430 research papers (till date) in journals and conferences, and 2 books are ongoing. He is in the editorial boards of various international journals, book series, and conferences. Prof. Chakraverty is the chief editor of the *International Journal of Fuzzy Computation and Modelling* (*IJFCM*), Inderscience Publisher, Switzerland (http://www.inderscience.com/ijfcm); the associate editor of *Computational Methods in Structural Engineering* and *Frontiers in Built Environment and Curved and Layered Structures* (De Gruyter); and the editorial board member of *Springer Nature Applied Sciences, IGI Research Insights Books*, Springer Book Series of *Modeling and Optimization in Science and Technologies, Coupled Systems Mechanics* (Techno Press), the *Journal of Composites Science* (MDPI), *Engineering Research Express* (IOP), and *Applications and Applied Mathematics: An International Journal*. He is also the reviewer of around 50 national and international journals of repute, and he was the president of the Section of Mathematical Sciences (including Statistics) of Indian Science Congress (2015–2016) and was the vice president—Orissa Mathematical Society (2011–2013). Prof. Chakraverty is a recipient of prestigious awards, namely the Indian National Science Academy (INSA) nomination under International Collaboration/Bilateral Exchange Program (with the Czech Republic), Platinum Jubilee ISCA Lecture Award (2014), CSIR Young Scientist Award (1997), BOYSCAST Fellowship (DST), UCOST Young Scientist Award (2007, 2008), Golden Jubilee Director's (CBRI) Award (2001), INSA International Bilateral Exchange Award ([2010–2011] selected but could not undertake [2015] selected), Roorkee University Gold Medals (1987, 1988) for first positions in M.Sc. and M.Phil. (Computer Application). He is in the list of 2% of world scientists (2020 and 2021) in Artificial Intelligence and Image Processing category based on an independent study done by Stanford University scientists. Prof. Chakraverty has received an IOP Publishing Top-Cited Paper Award in 2021 and 2022 from India, published across the entire IOP Publishing journal portfolio in the past three years, i.e., 2018–2020 and 2019–2021, respectively, for his paper that ranked among the top 1% of the most-cited papers in the materials subject category. In 2020–2021, one of his papers published in *ZAMM—Journal of Applied Mathematics and Mechanic* (Wiley) was among the most-cited papers. He has already guided 25 Ph.D. students and 12 are ongoing. Prof. Chakraverty has undertaken around 16 research projects as a principal investigator funded by international and national agencies totaling about Rs. 1.5 crores. He has boosted around eight international students with different international/national fellowships to work in his group as PDF, Ph.D., visiting researchers for different periods. A good number of international and national conferences, workshops, and training programs have also been organized by him. His present research area includes differential equations (ordinary, partial, and fractional), numerical analysis and computational methods, structural dynamics (FGM, nano) and fluid dynamics, mathematical and uncertainty modeling, soft computing, and machine intelligence (artificial neural network, fuzzy, interval, and affine computations).

Preface

The incorporation of uncertainties within the material properties of structural constituents has evolved into a prominent concern within the domain of structural analysis. Conventionally, deterministic or crisp parameters have been employed for structural analysis endeavors. However, the underlying reality underscores that the primary sources engendering uncertainty or randomness are intrinsic defects existing within atomic configurations, the presence of measurement errors, prevailing environmental conditions, and diverse contributory factors collectively that may influence the dynamic behaviors of structures. These structural aberrations denote that materials may not always exhibit their expected mechanical behavior. Regarding these concerns, it is necessary to propose new models that specifically integrate and effectively overcome imprecisely defined parameters of the system. With a focus on providing a timely contribution to this challenging field, in this book, uncertainty modeling of micro- or nano- and FG beams is presented using non-probabilistic approaches which include interval and fuzzy concepts. Vibration and buckling analyses of the beams are conducted using different analytical, semi-analytical, and numerical methods for finding exact and/or approximate solutions of governing equations arising in uncertain environments. Serving as a valuable resource for researchers, undergraduate and graduate students, and industry professionals, this book provides a one-stop resource for exploring the vibration and buckling analyses of different types of beams in fuzzy uncertainty and presents a comprehensive analysis for propagation of uncertainty in natural frequencies and buckling loads.

The present work comprises a total of eight chapters, while **Chapter 1** is devoted to provide a comprehensive introduction to uncertainties and dynamical analysis of different types of beams, namely micro or nano and functionally graded beams. On the other hand, **Chapter 2** provides a comprehensive overview of fundamental concepts in fuzzy set theory, including definitions of fuzzy sets and different types of fuzzy numbers such as triangular, trapezoidal, and Gaussian fuzzy numbers. Additionally, the concept of double parametric form, which plays a crucial role in representing uncertain parameters, is introduced.

Chapter 3 delves into the modeling of an Euler–Bernoulli nanobeam using Eringen's nonlocal theory, incorporating material uncertainties related to mass density and Young's modulus. A novel double parametric Rayleigh–Ritz method is introduced to address these uncertainties, allowing for the investigation of vibration characteristics and analysis of uncertainty propagation in frequency parameters. The model is validated through convergence studies, and lower and upper bounds of frequency parameters are computed using the double parameter approach. The sensitivity of the models to uncertainties is demonstrated through Triangular Fuzzy Numbers. Comparative analyses with existing literature show robust agreement in special cases, highlighting the reliability of the proposed methodology. The uncertainty modeling and frequency parameter bounds provide valuable insights for engineering structure design and optimization, enhancing quality and performance. In **Chapter 4**, a non-probabilistic-approach-based Navier's method (NM) and Galerkin Weighted Residual Method (GWRM) have been proposed to investigate the buckling behavior of Euler–Bernoulli nonlocal beams under Eringen's nonlocal elasticity theory. The uncertainties in Young's modulus and diameter are modeled using Triangular Fuzzy Numbers (TFNs). Critical buckling loads are calculated for hinged-hinged, clamped-hinged, and clamped-clamped boundary conditions and compared with the deterministic model in special cases. The Monte Carlo Simulation Technique (MCST) is used to compute the critical buckling loads of uncertain systems. The critical buckling loads obtained from the uncertain model are verified with the Monte Carlo Simulation Technique (MCST) over time periods, demonstrating the efficacy, accuracy, and effectiveness of the proposed uncertain model. A comparative study is conducted to demonstrate the effectiveness of the methods with respect to time.

Chapter 5 explores uncertainty quantification in nanobeam vibration by integrating first-order shear deformation beam theory with Eringen's nonlocal elasticity theory. Material uncertainties are represented using Triangular Fuzzy Numbers. The governing equations for vibration are derived using the von Kármán hypothesis and Hamilton's principle, leading to closed-form solutions through Navier's method. The frequency parameters obtained are compared with those in previous literature, demonstrating a strong agreement in specific cases. A Monte Carlo Simulation Technique (MCST) is used to assess the natural frequencies of the nanobeam amid material uncertainties, providing insights into the system's response variability. The results are validated by comparing the natural frequencies derived from Navier's method with those obtained through MCST. This analysis highlights the efficacy, accuracy, and robustness of the proposed uncertain model in capturing the dynamic behavior of the nanobeam under varying material conditions. The computation of lower and upper bounds of frequency parameters is plotted using Triangular Fuzzy Numbers.

Chapter 6 examines the impact of material uncertainties on the stability of a Timoshenko nanobeam. The uncertainties are linked to the nanobeam's diameter, length, and Young's modulus using a special fuzzy number, the Symmetric Gaussian Fuzzy Number (SGFN). The stability analysis of the uncertain model is based on Timoshenko beam theory, Hamilton's principle, and double parametric form of fuzzy numbers. Eringen's elasticity theory addresses the nanobeam's small-scale effect, and Navier's method calculates the results for the lower and upper bound of buckling loads. The results are validated with deterministic models and Monte Carlo Simulation techniques, demonstrating a robust agreement. A parametric study investigates the fuzziness or spreads of buckling loads with respect to different uncertain parameters.

Chapter 7 discusses the impact of material uncertainties on the vibration characteristics of functionally graded (FG) beams. These uncertainties are linked to Young's modulus and material density of the metal constituent, which are encapsulated in Triangular Fuzzy Numbers (TFNs). The governing equations for vibrational behavior of the uncertain model are derived using Euler–Bernoulli beam theory and a dual parametric form of fuzzy numbers. The Rayleigh–Ritz method is used to compute the outcomes for the lower bound and upper bound of natural frequencies under various boundary conditions. A rigorous validation against the deterministic model is conducted to ensure the model's robustness and reliability.

A comprehensive parametric exploration is undertaken to explore the intricacies of fuzziness or spreads related to natural frequency and uncertain parameters, providing valuable insights into the vibrational domain. Lastly, **Chapter 8** explores the impact of geometrical uncertainties on the free vibration of Euler–Bernoulli functionally graded (FG) beams on a Winkler–Pasternak elastic foundation. The uncertainties are linked to the beam's length and thickness using the Symmetric Gaussian Fuzzy Number (SGFN). The governing equations of motion for the uncertain model are derived by combining the SGFN with Hamilton's principle and double parametric form of fuzzy numbers. The natural frequencies of the uncertain models are computed using Navier's approach for the hinged-hinged boundary condition and Hermite–Ritz approach for hinged-hinged, clamped-hinged, and clamped-clamped boundary conditions. The results show a strong agreement with existing research, and a comprehensive parametric analysis is conducted to investigate the fuzziness or spreads of natural frequencies in relation to uncertain parameters.

We trust that the consolidated, challenging, and comprehensive approaches in all of the chapters will benefit the readers for their future studies and research. Finally, we would also like to thank the CRC team for their assistance and support throughout this project.

Subrat Kumar Jena and S. Chakraverty

Acknowledgment

The first author, Dr. Subrat Kumar Jena, is immensely grateful to his family members, specifically Shri Ullash Chandra Jena, Shri Durga Prasad Jena, Shri Laxmidhara Jena, Smt. Urbasi Jena, Smt. Renu Bala Jena, Smt. Arati Jena and his sisters Jyotirmayee, Truptimayee, and Nirupama, for their unwavering love, constant motivation, unrelenting support, and blessings. Dr. Jena expresses deep indebtedness to his younger brother, Dr. Rajarama Mohan Jena, and sister-in-law, Dr. Sujata Swain, for their encouragement, love, and support. Furthermore, the first author would like to express his deep appreciation to his Ph.D. supervisor, Prof. S. Chakraverty, for his invaluable guidance and support throughout the early stages of his research. Finally, Dr. Jena would like to acknowledge the administration of the Indian Institute of Technology Delhi and his postdoc mentor, Prof. S. Pradyumn, for their assistance.

Prof. S. Chakraverty, the second author, expresses gratitude to his beloved parents, late Shri Birendra K. Chakraborty and the late Parul Chakraborty, for their blessings. He also thanks his wife, Shewli, and daughters, Shreyati and Susprihaa, for their support and inspiration during this project. The support of the NIT Rourkela administration is also gratefully acknowledged.

Together, the two authors extend sincere gratitude and acknowledgment to the reviewers for their invaluable feedback and appreciation during the development of the book proposal. They are immensely grateful to the entire team at CRC for their support, cooperation, and assistance, enabling the timely publication of this book. Finally, they express deep indebtedness to the authors and researchers cited in the bibliography/reference sections at the end of each chapter.

1 Introduction to Structural Uncertainties and Micro-, Nano-, and Functionally Graded Beams

The uncertainties or randomness of the material properties of structural components have become a serious concern in the field of structural analysis. Typically, deterministic or crisp parameters are used in structural analysis, but the truth is that the primary causes of uncertainty or randomness are defects in atomic configurations, measurement errors, environmental conditions, and other factors that hinder the behavior of dynamical structures. These structural anomalies indicate that materials may not have the ability to demonstrate their normal mechanical behaviors. Furthermore, the influence of uncertainties becomes much more profound in the case of nano- and microstructures due to the small-scale effects, as evidenced by several nanoscale experiments and molecular dynamics studies. For example, Salvetat et al. (1999) and Krishnan et al. (1998) experimentally observed significant errors in Young's modulus and shear modulus of carbon nanotubes, indicating inherent uncertainties and randomness. The study by He et al. (2014) investigated the effect of Stone–Thrower–Wales (STW) defects on the mechanical properties of graphene sheets through molecular dynamics simulations. The study revealed that STW defects significantly affected the mechanical properties of graphene sheets, including the elastic modulus, tensile strength, and fracture behavior. The results showed that STW defects caused a reduction in the elastic modulus and tensile strength of the graphene sheets, as well as a change in the fracture behavior from brittle to ductile. The findings provided insights into the effect of defects on the mechanical properties of graphene and had potential applications in the design and engineering of graphene-based materials. From these investigations, it is inevitable that the defects in atomic configurations and measurement errors have inherent uncertainties and randomness.

To address these concerns, it is necessary to propose new models that integrate and effectively overcome imprecisely defined parameters of the system. Over the last few decades, several factors have been introduced to ensure the safety protocol with respect to impreciseness. However, the need for substantive productivity and reliability, better performance, and reduced costs mandate the design and production of an improved system. In this context, the literature survey related to structural uncertainties and their propagation, nanobeams, functionally graded beams, functionally graded nanobeams, microbeams, and functionally graded microbeams with respect to buckling and vibration is discussed in the next section.

DOI: 10.1201/9781003303107-1

1.1   UNCERTAINTY IN STRUCTURAL ELEMENTS AND ITS PROPAGATION

To address uncertainties in modeling, three approaches can be utilized: Probabilistic, non-probabilistic (which includes interval and fuzzy), and Monte Carlo Simulation. Probabilistic methods require a large amount of experimental data due to the require-ment of the probability density function for the uncertain parameters as random variables. However, this method often does not yield reliable results within a toler-ated error range due to inadequate experimental data. As a result, non-probabilistic methods, including the interval and fuzzy approaches, are being increasingly applied as an effective means of dealing with uncertainty in modeling. Zadeh (1996) was the first to propose a fundamental concept in fuzzy theory. Later, using this concept, several uncertainty problems were solved, which can be found in Hanss and Turrin (2010), Khastan et al. (2011), Mikaeilvand and Khakrangin (2012), Khastan et al. (2013), and Tapaswini and Chakraverty (2014).

Qiu and Wang (2003) conducted a comparison study on the dynamic response of structures with uncertain-but-bounded parameters using the non-probabilistic interval analysis method and the probabilistic approach. The results showed that the non-proba-bilistic approach offers a more conservative estimate of the response, which can be use-ful in situations where limited information is available about uncertain parameters. Gao et al. (2011) developed a hybrid probabilistic interval analysis method for bar structures with uncertainty. They employed a mixed perturbation Monte Carlo method to generate random samples and applied interval analysis to obtain bounds on the response. The study demonstrated that the method is accurate and efficient, making it suitable for complex systems. Muscolino and Sofi (2012) presented an improved interval analysis method for stochastic analysis of structures with uncertain-but-bounded parameters. The proposed algorithm aims to improve the accuracy and efficiency of the response estimates by handling the bounds of the uncertain parameters in a more effective way. Xu et al. (2015) investigated the stochastic dynamic characteristics of function-ally graded material (FGM) beams with random material properties. They employed a probabilistic approach based on the stochastic finite element method to model the mate-rial properties and then analyzed the dynamic response of the beams. The study found that material randomness significantly affects the dynamic behavior of FGM beams.

Liu et al. (2017) conducted a study on flexural wave propagation in fluid-convey-ing carbon nanotubes with system uncertainties. They used the polynomial chaos expansion method to model the uncertainty in the system parameters and analyzed the propagation of waves in the nanotubes. The study found that the fluid viscosity and system uncertainties significantly affect wave dispersion and attenuation. Lv and Liu (2017) explored the nonlinear bending response of functionally graded nano-beams with material uncertainties, represented as interval parameters, and proposed an iterative algorithm-based interval analysis method to solve the model. In another innovative study, Lv and Liu (2018) investigated the influence of nanomaterial uncer-tainties on the vibration and buckling characteristics of functionally graded nano-beams in a thermal environment. They established a non-probabilistic uncertainty modeling approach for FG nanobeams by quantifying nanomaterial uncertainties as interval parameters and developed a hull iterative method to solve the model. The effect of material uncertainty on wave propagation in nanorods placed in elastic substrates has been thoroughly investigated by Lv et al. (2018). In a related study, Liu and Lv (2019) examined the vibration characteristics of a magneto-electro-elastic nanobeam embedded in a Winkler–Pasternak elastic foundation, considering material uncertainties in interval form.

1.2 NANOBEAMS

Nanostructures have been the focus of much scientific and engineering research due to their unique mechanical, electrical, and electronic properties. These characteristics make nanomaterials (Chakraverty and Behera, 2016), such as nanowires, nanoparticles, nanoribbons, and nanotubes, invaluable components of various nano-electromechanical systems, including nanoprobes, nanotube resonators, and nanoactuators. As such, extensive research is being conducted globally to better understand the mechanical properties of nanostructures, specifically nanobeams, to make accurate predictions about their dynamical characteristics. There are three approaches to exploring the mechanical behaviors of structural elements: Atomistic, semi-continuum, and continuum, with the latter being classified as classical or nonclassical. The classical continuum approach disregards the lattice gap between individual atoms and ignores small-scale effects induced by the electric force, chemical bond, and van der Waals forces. However, the experimental and atomistic simulation results demonstrate that ignoring the nanoscale effect is not viable. Hence, other nonclassical continuum approaches have emerged, such as strain gradient theory, couple stress theory, modified couple stress theory, micropolar theory, and nonlocal elasticity theory.

Conducting experiments at the nanoscale is a challenging and resource-intensive undertaking. Therefore, developing appropriate mathematical models for the dynamical characteristics of nanobeams is essential for their practical applications. The small-scale effect, a crucial factor in nonclassical continuum approaches, has been calibrated using molecular dynamics simulations (Duan et al., 2007; Huang et al., 2012). When the size of the beam is reduced to the nanoscale, nonlocal effects become significant for predicting natural frequencies and vibrating modes, especially higher-order natural frequencies and vibrating modes (Xu, 2006). Nonlocal effects play a crucial role in nanoscale devices (Peddieson et al., 2003). Wang et al. (2006) investigated the buckling behavior of micro- and nanorods/tubes using Timoshenko beam theory and Eringen's nonlocal elasticity theory. Analytical solutions for the vibration of nonlocal Timoshenko nanobeams were studied by Wang et al. (2007) using nonlocal elasticity theory. Reddy (2007) and Aydogdu (2009) analyzed the buckling, bending, and vibration of Euler–Bernoulli, Timoshenko, Reddy, and Levinson beam theories analytically, using the nonlocal differential version of Eringen's nonlocal elasticity theory.

Emam (2013) utilized nonlocal elasticity theory to analyze the buckling and post-buckling response of nanobeams. Yu et al. (2016) studied the buckling behavior of

Euler–Bernoulli nanobeams with a nonuniform temperature distribution using nonlocal thermoelasticity theory. Additionally, Li et al. (2016) explored the longitudinal vibration of rods using a nonlocal strain gradient model with an analytical approach and a finite element method. Barretta et al. (2016) investigated the higher-order version of Eringen's model using the Euler–Bernoulli beam, while Eptaimeros et al. (2016) studied the implementation of the integral form of nonlocal Euler–Bernoulli beam for dynamic response analysis. Khaniki and Hosseini-Hashemi (2017) employed the generalized differential quadrature method to investigate the buckling behavior of Euler–Bernoulli beams with various types of cross-section variation. Rahimi et al. (2017) used conformable fractional derivative with Galerkin weighted residual method to study the free vibration of Euler–Bernoulli and Timoshenko nanobeams for simply supported boundary conditions. Dai et al. (2018) examined the pre- and post-buckling behavior of nonlocal nanobeams when subjected to axial and longitudinal magnetic forces analytically, while Yu et al. (2019) proposed a size-dependent gradient-beam model with three characteristic lengths by incorporating modified nonlocal theory and Euler–Bernoulli beam model.

Malikan et al. (2019) investigated the transient response of a nanotube with a simply supported boundary condition using the Kelvin–Voigt viscoelasticity model in conjunction with the nonlocal strain gradient theory. Based on the nonlocal strain gradient theory and incorporating surface effects, Esfahani et al. (2019) investigated the size-dependent nonlinear vibration of an electrostatic nanobeam actuator, while Jankowski et al. (2020) explored the effect of porous material on bifurcation buckling and natural vibrations of nanobeams by incorporating higher-order nonlocal strain gradient theory and Reddy beam theory. Hamidi et al. (2020) studied the torsional vibrations of nanobeams under distributed external torque and moving external harmonic torque based on the nonlocal strain gradient theory and surface effects. Ebrahimi and Mahesh (2021) investigated the damping forced harmonic vibration characteristics of magneto-electro-viscoelastic nanobeams embedded in Winkler–Pasternak foundations. Jena et al. (2022c) proposed a novel numerical approach for analyzing the stability of nanobeams under the influence of hygro-thermo-magnetic environmental conditions and embedded in an elastic foundation. The results demonstrated the efficacy of the proposed approach in analyzing the stability of nanobeams under the influence of multiple environmental factors, which was of significance for the design and performance of nanomechanical devices. Ahmad et al. (2022) investigated the thermoelastic vibrations of Euler–Bernoulli nanobeams using Eringen's nonlocal elasticity theory and the generalized thermoelastic model with dual phase-lag to investigate the effect of temperature change. In another work, hygro-magneto-thermo vibration behavior of nonlocal strain gradient nanobeams resting on a Winkler–Pasternak elastic foundation was investigated by Jena et al. (2022b) by adopting wavelet-based two relatively recent techniques, namely the Haar Wavelet Method (HWM) and the Higher Order Haar Wavelet Method (HOHWM). The effects of several scaling parameters on frequency parameters are thoroughly investigated for pined-pined and clamped-clamped boundary conditions. Karamanli and Vo (2022) utilized a finite element doublet mechanics model to investigate free vibrations of curved zigzag nanobeams using sinusoidal beam theory and doublet mechanics formulation.

1.3 FUNCTIONALLY GRADED BEAMS

The concept of functionally graded materials (FGMs) was first introduced in 1984 by a group of material scientists in Japan (Koizumu, 1993), who intended to use it in a space aircraft project as a thermal barrier material that can withstand high-temperature variations over a very thin cross-section. Since then, FGMs have gained substantial attention as advanced heat-shielding structural materials in various engineering applications and industrial sectors, including aerospace, nuclear power, automobiles, aviation, space vehicles, biomedical, and steel (Chakraverty and Pradhan, 2016, 2018). FGMs are typically inhomogeneous materials that comprise ceramic-metal composites, and the composition or volume of constituents varies continuously along one or more specific directions. As a result, their properties vary steadily from one interface to the next in a predetermined mathematical pattern. The ceramic component has low thermal conductivity, making it suitable for use in high-temperature environments such as nuclear reactors and chemical plants and in the manufacture of high-speed vessels. On the other hand, the ductile metal component avoids fracture caused by strains induced by a high-temperature gradient. Consequently, the mechanical strength of FGMs can be significantly enhanced while reducing the weight of the structure, resulting in cost savings.

Sina et al. (2009) conducted an analytical investigation on the free vibration of shear deformable FGM beams, exploring various boundary conditions. Ke et al. (2010) focused on the nonlinear vibration of FGM beams, utilizing Euler–Bernoulli beam theory and von Kármán geometric nonlinearity for hinged-hinged, clamped-hinged, and clamped-clamped boundary conditions. The Galerkin procedure and Runge–Kutta method were utilized to solve the governing equations. Hein and Feklistova (2011) applied the Haar wavelet method to the study of nonuniform functionally graded beams with varying cross-sections and different boundary conditions, employing the Euler–Bernoulli theory and assuming material properties varied along the axial direction. Shooshtari and Rafiee (2011) developed a nonlinear Euler–Bernoulli FGM beam model to investigate the free and forced vibrations of FGM beams with clamped ends using von Kármán nonlinearity, considering exponentially or power-law distributed material characteristics. Wattanasakulpong et al. (2011) explored the thermal buckling and vibration of functionally graded beams using an improved third-order shear deformation theory for clamped boundary conditions and the Ritz technique. Wattanasakulpong et al. (2012) examined the free vibration of layered functionally graded beams with various boundary conditions based on an improved third-order shear deformation theory and using the Ritz technique to verify the frequencies by theoretical and experimental results. Additionally, Thai and Vo (2012) developed various higher-order shear deformation beam theories for bending and free vibration of functionally graded beams, assuming the material properties of the functionally graded beam varied according to the power-law distribution of the volume fraction. Hamilton's principle was used to obtain the governing equations, and the Navier solution procedure was applied to solve them. Pradhan and Chakraverty (2013) used the Rayleigh–Ritz method to explore the free vibration of Euler and Timoshenko functionally graded beams subjected to various boundary conditions. They also examined the effects of constituent volume fractions, slenderness ratios, and beam theories on the natural frequencies. Rahimi et al. (2013) employed von-Kármán-type nonlinear strain–displacement equations to investigate post-buckling and free vibrations of FG Timoshenko beams with fixed–fixed, fixed–hinged, and hinged–hinged boundary conditions. Based on a refined shear deformation theory, Vo et al. (2014) proposed a finite element model for the vibration and buckling of functionally graded sandwich beams. Hamilton's principle generated the governing equations of motion and boundary conditions. Kanani et al. (2014) employed the Variational Iteration Method to investigate the free and forced vibration of a functionally graded beam resting on a nonlinear elastic substrate. They generated the equations of motion using Euler–Bernoulli beam theory and von Kármán geometric nonlinearity. Su and Banerjee (2015) employed the dynamic stiffness approach to investigate the free vibration of functionally graded Timoshenko beams, and they yielded the natural frequencies using the Wittrick–Williams algorithm technique. Tossapanon and Wattanasakulpong (2016) analyzed buckling and vibration problems of functionally graded sandwich beams resting on a two-parameter elastic foundation using the Timoshenko beam theory and the Chebyshev collocation approach. Akbaş (2017) used the finite element approach to study the temperature effects on the free vibration of functionally graded porous deep beams. They discussed the effects of porosity parameters, material distribution, porosity models, and temperature rise on the vibration characteristics. Avcar and Mohammed (2018) investigated the free vibration of functionally graded beams resting on a two-parameter elastic substrate using classical beam theory. The foundation medium was assumed to be linear, homogeneous, and isotropic, and the reaction of the elastic foundation on the beam was described using the Winkler–Pasternak model. Finally, Celebi et al. (2018) investigated the free vibration of simply supported functionally graded beams using the complementary functions approach and plane elasticity theory with modulus of elasticity, density of material, and Poisson's ratio varied arbitrarily in the thickness direction.

Sinir et al. (2018) utilized the differential quadrature technique to study the nonlinear free and forced vibrations of axially functionally graded beams with nonuniform cross-sections, subject to clamped–clamped and pinned–pinned boundary conditions, based on the Euler–Bernoulli beam theory. Banerjee and Ananthapuvirajah (2018) employed the dynamic stiffness method to explore the free vibration of a functionally graded beam, considering the power-law exponent model to describe the variation of material properties along the thickness direction. Cao and Gao (2019) investigated the free vibration of nonuniform axially functionally graded beams, using an asymptotic development technique to obtain an approximate analytical expression for the natural frequencies under various boundary conditions. Ma et al. (2020) proposed a refined beam theory or third-order shear deformation beam theory to study transverse bending and vibration of functionally graded circular cylindrical tubes with radial nonhomogeneity, accounting for warping, shear deformation, and rotational moment of inertia of the cross-section under clamped–clamped, pinned–pinned, and clamped-free boundary conditions. Selmi (2020) presented an exact solution for the behavior of clamped–clamped functionally graded buckling beams, taking into account von Kármán geometric nonlinearity and varying effective material characteristics according to an exponential law along the thickness direction. Jena et al. (2021) explored the free vibration of a functionally graded (FG) beam with uniformly distributed porosity along the beam's thickness using the shifted Chebyshev polynomial-based Rayleigh–Ritz method and Navier's technique. According to the power-law exponent model, the material properties such as Young's modulus, mass density, and Poisson's ratio were therefore considered to vary along the thickness of the FG beam. The porous FG beam was indeed embedded in the Kerr elastic foundation, and the displacement field of the beam is governed by a refined higher-order shear deformation theory. Based on 2-D elastic theory, Li et al. (2021) investigated the free vibration of functionally graded beams with different cross-sections resting on Pasternak elastic foundations using a variable method, Laplace transform, and Fourier series expansion. These studies have contributed to a deeper understanding of the dynamic behavior of functionally graded beams under different conditions and have potential implications for the design and analysis of various engineering structures.

1.4   FUNCTIONALLY GRADED NANOBEAMS

The integration of functionally graded materials into nanotechnology has led to the development of innovative devices and equipment that exhibit enhanced performance, including nano-electro-mechanical systems, thin shape memory alloys, atomic force microscopy, and more. Nanotechnology is the study of microscopic objects with sizes ranging from 1 to 100 nanometers, and its applications span various disciplines such as chemistry, biology, physics, and materials science. Given the distinctive mechanical properties of nanomaterials, researchers have increasingly investigated the use of nanoscale structures to build high-performance instruments like nanosensors, nano actuators, and nanogenerators to address new challenges. Notably, experimental studies and molecular simulations have demonstrated that nanomaterials exhibit distinct responses at the nanoscale compared to the macroscale, indicating that size is a critical factor in the nanoscale. To predict the mechanical response of these materials, nonclassical continuum elasticity theories have proven to be among the most efficient and cost-effective techniques. Classical continuum theories have been found to be inadequate in representing the behavior of nanoscale structures, as they do not account for size effects. As such, nonlocal continuum theories have been developed to address these shortcomings. The most prominent nonlocal continuum theories include strain gradient theory (Mindlin, 1965), nonlocal elasticity theory (Eringen, 2002), stress-driven nonlocal elasticity theory (Barretta et al., 2018), nonlocal strain gradient theory (Lim et al., 2015), modified coupled stress theory (Yang et al., 2002), surface elasticity theory (Ansari et al., 2013), and bi-Helmholtz nonlocal elasticity theory (Lazar et al., 2006; Koutsoumaris and Eptaimeros, 2018). It should be noted that each of the aforementioned theories have a small-scale parameter or length-scale parameter. Many studies have found that these parameters are not material constants and vary with the intrinsic characteristics and physical properties of nanomaterials. Furthermore, since nanomaterials are temperature-dependent, the thermal environment may significantly impact the value of the small-scale parameter. As a result, distinct values of the small-scale parameter are required to produce accurate results for nanostructures under varied boundary conditions and external temperatures. Additionally, other parameters, such as crack specifications in cracked nanomaterials and atomic lattice arrangement in special nanomaterials like graphene and nanotubes with configurable arrangements (chirality effect), may influence the small-scale parameter.

Over the past decade, there has been a growing interest among researchers in investigating the dynamical behavior of functionally graded nanomaterials with different geometries and boundary conditions due to their widespread applications. In particular, several studies have been published on the dynamics of functionally graded nanobeams. Eltaher et al. (2012) analyzed the free vibration of a functionally graded nanobeam based on Euler–Bernoulli beam theory, with the nonlocal elasticity theory of Eringen used to capture the small-scale effect of the beam. The numerical outputs were obtained for various boundary conditions using the finite element method. Sharabiani and Yazdi (2013) investigated the nonlinear free vibration of functionally graded nanobeams using Euler–Bernoulli beam theory and von Kármán geometric nonlinearity, while also exploring the effects of surface elasticity on the nonlinear natural frequencies of a functionally graded nanobeam. Nazemnezhad and Hosseini-Hashemi (2014) examined the nonlinear free vibration of functionally graded nanobeams using Euler–Bernoulli beam theory and von Kármán nonlinearity. They utilized a multiple-scale technique to obtain an analytical solution for natural frequencies for SS and SC boundary conditions. Hosseini-Hashemi et al. (2014) investigated the effects of surface elasticity, surface tension, surface density, and nonlocal parameter on the nonlinear free vibration of functionally graded nanobeams using the Euler–Bernoulli beam theory with von Kármán geometric nonlinearity. Ansari et al. (2015) explored the effects of surface elasticity and surface stress on the forced vibration response of FG nanobeams under thermal stresses. They utilized the Gurtin–Murdoch elasticity theory to develop a nonclassical beam model, and the governing equation of motion and accompanying boundary conditions were established using Hamilton's principle. The Galerkin and various time scales perturbation techniques were used to study the principal resonance phenomena. Zeighampour and Tadi Beni (2015) investigated the vibration of axially functionally graded material nanobeams using the differential quadrature technique, with the nanoscale effect captured by the strain gradient theory. The nanobeam was modeled as a visco-Pasternak foundation, and Hamilton's principle and Euler–Bernoulli beam theory were used to derive the governing equations and boundary conditions. Ebrahimi and Salari (2015) employed a Navier-type solution and differential transform method to examine the thermal influence on the free vibration characteristics of functionally graded size-dependent nanobeams exposed to various types of thermal loading. Şimşek (2016) proposed a new size-dependent beam model for nonlinear free vibration of a functionally graded nanobeam with immovable ends based on the nonlocal strain gradient theory and Euler–Bernoulli beam theory in combination with von-Kármán's geometric nonlinearity. Shafiei et al. (2016a) investigated the nonlinear vibration of axially functionally graded nonuniform nanobeams using the generalized differential quadrature and homotopy perturbation methods based on Eringen's nonlocal elasticity and the Euler–Bernoulli beam model with von-Kármán's geometric nonlinearity. Arefi and Zenkour (2017) used Timoshenko's beam and nonlocal elasticity theories to investigate wave propagation for a functionally graded nanobeam with rectangular cross-section resting on visco-Pasternak's elastic matrix. Mirjavadi et al. (2018) investigated buckling and nonlinear vibration of functionally graded porous nanobeams using the generalized differential quadrature method and nonlinear Von Kármán strains in conjunction with Euler–Bernoulli beam theory for clamped–clamped and simply supported–simply supported boundary conditions. Şimşek (2019) explored static bending, buckling, and free and forced vibration of functionally graded nanobeams analytically using nonlocal strain gradient theory and Euler–Bernoulli beam theory. Aria and Friswell (2019) used a nonlocal finite element approach based on the first-order shear deformation theory to investigate the free vibration and buckling behavior of functionally graded nanobeams. Uzun and Yayli (2019) investigated the free vibration of functionally graded nanobeams under hinged–hinged and clamped–clamped boundary conditions while considering the nonlocal impact of the FG nanobeam using Eringen's nonlocal theory with the use of a finite element model. Esmaeili and Tadi Beni (2019) investigated the buckling and vibration characteristics of a flexoelectric smart nanobeam composed of functionally graded materials for simply supported and clamped boundary conditions using Euler–Bernoulli beam theory and von Kármán strain. Khaniki (2019) used a modified differential quadrature approach to investigate the vibration behavior of axially functionally graded nanobeams in the framework of Eringen's two-phase local-nonlocal model. Uzun and Yayli (2020) used a combination of Euler–Bernoulli beam theory and Eringen's nonlocal elasticity to investigate the free vibration of functionally graded nanobeams for simply supported boundary conditions using a finite element method. Jena et al. (2020) studied the vibrational properties of a functionally graded porous nanobeam embedded in a Winkler–Pasternak-type elastic substrate. The study utilized classical beam theory, bi-Helmholtz type nonlocal elasticity, and the Hermite–Ritz method to compute natural frequencies under different boundary conditions. Using Navier's approach, a closed-form solution was obtained for a hinged–hinged boundary condition. The study also emphasized the benefits of utilizing Hermite polynomials as shape functions. Finally, Chen et al. (2020) analytically investigated the thermal buckling of an Euler–Bernoulli beam composed of FG material using the transformed-section approach.

1.5 MICROBEAMS

The study of structural elements is crucial for the effective investigation of micro- or nano-electromechanical systems, including nanosensors, nanoactuators, nano-motors, and atomic force microscopes (de Souza Pereira, 2001; Pei et al., 2004). Size-dependent effects have a significant impact due to the small size of a beam and are not accurately captured by classical elasticity theory (Lam et al., 2003; McFarland and Colton, 2005). Nonlocal continuum theories, which contain additional material length-scale parameters, have been proposed to account for the microstructure-dependent size effect. The classical couple stress theory was first proposed by Mindlin (1962), Toupin (1962), and Mindlin and Tiersten (1962) for the isotropic elastic material with four material constants. However, assessing the microstructure-related length-scale parameters can be complex (Yang and Lakes, 1982). To address this, Yang et al. (2002) introduced a modified version of the couple stress theory with only one material length-scale parameter. Park and Gao (2006) proposed a new model for bending of Bernoulli–Euler beam employing the modified couple stress theory and applied it to a cantilever beam problem. In the work of Park and Gao (2008), a variational formulation for the modified couple stress theory was proposed, leading to the simultaneous determination of the equilibrium equations and the boundary conditions. Asghari et al. (2010) employed the modified couple stress theory to investigate the bending and free vibration analysis of a nonlinear size-dependent Timoshenko beam using a multiple scales technique. Jam et al. (2017) used the Galerkin method to study the nonlinear free vibration analysis of microbeams resting on the viscoelastic foundation, employing the modified couple stress theory. Using the modified couple stress theory, Hieu (2018) investigated the post-buckling and nonlinear free vibration of microbeams resting on a nonlinear elastic foundation. The governing equation of motion was solved using Galerkin's method for pinned–pinned and clamped–clamped boundary conditions.

1.6   FUNCTIONALLY GRADED MICROBEAMS

A significant amount of research has been carried out to analyze the vibration characteristics of micro and nano-systems, given their widespread use in various applications. The development of composites, a class of advanced materials composed of one or more materials that exhibit distinct physical and chemical properties when combined in a solid state, has emerged in response to the demand for materials with contrasting properties such as being lightweight and having high strength, hardness, and ductility (Mahamood et al., 2012). Delamination, also known as interlaminar cracking, is a primary concern associated with composites, and it is one of the most common types of composite damage (Wang, 1983). Functionally graded materials (FGMs) were developed to address this issue and improve performance. FGMs are composite materials consisting of two or more components with varying characteristics, and their volume fractions vary smoothly between surfaces. As a result of these smooth variations, their mechanical and physical characteristics vary constantly in a particular direction. The most crucial feature of FGMs is the continuous modification of their properties, which eliminates the existence of sharp interfaces in composite materials, where failure often occurs. FGMs find extensive use in various industries, such as aerospace, medicine, military, energy, and optoelectronics.

Ke and Wang (2011) investigated the dynamic stability of functionally graded microbeams utilizing modified couple stress theory and Timoshenko beam theory. The material characteristics were estimated using the Mori–Tanaka homogenization approach, which was assumed to vary in the thickness direction. Şimşek and Reddy (2013) investigated the bending and free vibration of an FGM microbeam utilizing modified couple stress theory and several higher-order beam theories based on Navier's approach. Nateghi and Salamat-Talab (2013) investigated the thermal effect on size-dependent behavior of FG microbeams using classical and first-order shear deformation beam theories, with the modified couple stress theory being utilized. Shafiei et al. (2016b) carried out a significant investigation of the size-dependent vibration behavior of a rotating nonuniform functionally graded (FG) Timoshenko and Euler–Bernoulli microbeam using the generalized differential quadrature element technique and modified couple stress theory. The authors also examined the effect of shear deformation on the natural frequencies of the microbeam using different values of the material length-scale parameter, the rotational velocity, and the rate of cross-section change. Jena et al. (2022a) utilized Haar Wavelet Discretization Method (HWDM) and Differential Quadrature Method (DQM) to analyze the free vibration of a functionally graded (FG) microbeam with porosity distributed uniformly along its thickness. The effect of the power-law exponent, the porosity volume fraction index, and the thickness to material length-scale parameter on the natural frequencies was also explored. This concludes the concise literature review on structural uncertainties and their propagation, focusing on nanobeams, functionally graded beams, functionally graded nanobeams, microbeams, and functionally graded microbeams in the context of buckling and vibration analysis.

BIBLIOGRAPHY

Ahmad, H., Abouelregal, A.E., Benhamed, M., Alotaibi, M.F. and Jendoubi, A., 2022. Vibration analysis of nanobeams subjected to gradient-type heating due to a static magnetic field under the theory of nonlocal elasticity. *Scientific Reports*, *12*(1), p. 1894.

Akbaş, Ş.D., 2017. Thermal effects on the vibration of functionally graded deep beams with porosity. *International Journal of Applied Mechanics*, *9*(5), p. 1750076.

Ansari, R., Mohammadi, V., Shojaei, M.F., Gholami, R. and Sahmani, S., 2013. Postbuckling characteristics of nanobeams based on the surface elasticity theory. *Composites Part B: Engineering*, *55*, pp. 240–246.

Ansari, R., Pourashraf, T. and Gholami, R., 2015. An exact solution for the nonlinear forced vibration of functionally graded nanobeams in thermal environment based on surface elasticity theory. *Thin-Walled Structures*, *93*, pp. 169–176.

Arefi, M. and Zenkour, A.M., 2017. Analysis of wave propagation in a functionally graded nanobeam resting on visco-Pasternak's foundation. *Theoretical and Applied Mechanics Letters*, *7*(3), pp. 145–151.

Aria, A.I. and Friswell, M.I., 2019. A nonlocal finite element model for buckling and vibration of functionally graded nanobeams. *Composites Part B: Engineering*, *166*, pp. 233–246.

Asghari, M., Kahrobaiyan, M.H. and Ahmadian, M., 2010. A nonlinear Timoshenko beam formulation based on the modified couple stress theory. *International Journal of Engineering Science*, *48*(12), pp. 1749–1761.

Avcar, M. and Mohammed, W.K.M., 2018. Free vibration of functionally graded beams resting on Winkler-Pasternak foundation. *Arabian Journal of Geosciences*, *11*(10), pp. 1–8.

Aydogdu, M., 2009. A general nonlocal beam theory: Its application to nanobeam bending, buckling and vibration. *Physica E: Low-Dimensional Systems and Nanostructures*, *41*(9), pp. 1651–1655.

Banerjee, J.R. and Ananthapuvirajah, A., 2018. Free vibration of functionally graded beams and frameworks using the dynamic stiffness method. *Journal of Sound and Vibration*, *422*, pp. 34–47.

Barretta, R., Čanadija, M. and Marotti de Sciarra, F., 2016. A higher-order Eringen model for Bernoulli–Euler nanobeams. *Archive of Applied Mechanics*, *86*(3), pp. 483–495.

Barretta, R., Fabbrocino, F., Luciano, R. and de Sciarra, F.M., 2018. Closed-form solutions in stress-driven two-phase integral elasticity for bending of functionally graded nanobeams. *Physica E: Low-Dimensional Systems and Nanostructures*, *97*, pp. 13–30.

Cao, D. and Gao, Y., 2019. Free vibration of nonuniform axially functionally graded beams using the asymptotic development method. *Applied Mathematics and Mechanics*, *40*(1), pp. 85–96.

Celebi, K., Yarimpabuc, D. and Tutuncu, N., 2018. Free vibration analysis of functionally graded beams using complementary functions method. *Archive of Applied Mechanics*, *88*(5), pp. 729–739.

Chakraverty, S. and Behera, L., 2016. *Static and dynamic problems of nanobeams and nanoplates*. World Scientific.

Chakraverty, S. and Pradhan, K.K., 2016. *Vibration of functionally graded beams and plates*. Academic Press.

Chakraverty, S. and Pradhan, K.K., 2018. *Computational structural mechanics: Static and dynamic behaviors*. Academic Press.

Chen, W.R., Chen, C.S. and Chang, H., 2020. Thermal buckling analysis of functionally graded Euler-Bernoulli beams with temperature-dependent properties. *Journal of Applied and Computational Mechanics*, *6*(3), pp. 457–470.

Dai, H.L., Ceballes, S., Abdelkefi, A., Hong, Y.Z. and Wang, L., 2018. Exact modes for post-buckling characteristics of nonlocal nanobeams in a longitudinal magnetic field. *Applied Mathematical Modelling*, *55*, pp. 758–775.

de Souza Pereira, R., 2001. Atomic force microscopy as a novel pharmacological tool. *Biochemical Pharmacology*, *62*(8), pp. 975–983.

Duan, W.H., Wang, C.M. and Zhang, Y.Y., 2007. Calibration of nonlocal scaling effect parameter for free vibration of carbon nanotubes by molecular dynamics. *Journal of applied physics*, *101*(2), p. 024305.

Ebrahimi, F. and Mahesh, V., 2021. Chaotic dynamics and forced harmonic vibration analysis of magneto-electro-viscoelastic multiscale composite nanobeam. *Engineering with Computers*, *37*(2), pp. 937–950.

Ebrahimi, F. and Salari, E., 2015. Nonlocal thermo-mechanical vibration analysis of functionally graded nanobeams in thermal environment. *Acta Astronautica*, *113*, pp. 29–50.

Eltaher, M.A., Emam, S.A. and Mahmoud, F.F., 2012. Free vibration analysis of functionally graded size-dependent nanobeams. *Applied Mathematics and Computation*, *218*(14), pp. 7406–7420.

Emam, S.A., 2013. A general nonlocal nonlinear model for buckling of nanobeams. *Applied Mathematical Modelling*, *37*(10–11), pp. 6929–6939.

Eptaimeros, K.G., Koutsoumaris, C.C. and Tsamasphyros, G.J., 2016. Nonlocal integral approach to the dynamical response of nanobeams. *International Journal of Mechanical Sciences*, *115*, pp. 68–80.

Eringen, A.C., 2002. *Nonlocal continuum field theories*. Springer, Berlin.

Esfahani, S., Esmaeilzade Khadem, S. and Ebrahimi Mamaghani, A., 2019. Size-dependent nonlinear vibration of an electrostatic nanobeam actuator considering surface effects and inter-molecular interactions. *International Journal of Mechanics and Materials in Design*, *15*(3), pp. 489–505.

Esmaeili, M. and Beni, Y.T., 2019. Vibration and buckling analysis of functionally graded flexoelectric smart beam. *Journal of Applied and Computational Mechanics*, *5*(5), pp. 900–917.

Gao, W., Wu, D., Song, C., Tin-Loi, F. and Li, X., 2011. Hybrid probabilistic interval analysis of bar structures with uncertainty using a mixed perturbation Monte-Carlo method. *Finite Elements in Analysis and Design*, *47*(7), pp. 643–652.

Hamidi, B.A., Hosseini, S.A. and Hayati, H., 2020. Forced torsional vibration of nanobeam via nonlocal strain gradient theory and surface energy effects under moving harmonic torque. *Waves in Random and Complex Media*, pp. 1–16.

Hanss, M. and Turrin, S., 2010. A fuzzy-based approach to comprehensive modeling and analysis of systems with epistemic uncertainties. *Structural Safety*, *32*(6), pp. 433–441.

He, L., Guo, S., Lei, J., Sha, Z. and Liu, Z., 2014. The effect of Stone—Thrower—Wales defects on mechanical properties of graphene sheets—A molecular dynamics study. *Carbon*, *75*, pp. 124–132.

Hein, H. and Feklistova, L., 2011. Computationally efficient delamination detection in composite beams using Haar wavelets. *Mechanical Systems and Signal Processing*, *25*(6), pp. 2257–2270.

Hieu, D.V., 2018. Postbuckling and free nonlinear vibration of microbeams based on nonlinear elastic foundation. *Mathematical Problems in Engineering*, *2018*.

Hosseini-Hashemi, S., Nazemnezhad, R. and Bedroud, M., 2014. Surface effects on nonlinear free vibration of functionally graded nanobeams using nonlocal elasticity. *Applied Mathematical Modelling*, *38*(14), pp. 3538–3553.

Huang, L.Y., Han, Q. and Liang, Y.J., 2012. Calibration of nonlocal scale effect parameter for bending single-layered graphene sheet under molecular dynamics. *Nano*, *7*(5), p. 1250033.

Jam, J.E., Noorabadi, M. and Namdaran, N., 2017. Nonlinear free vibration analysis of microbeams resting on viscoelastic foundation based on the modified couple stress theory. *Archive of Mechanical Engineering*, *64*(2).

Jankowski, P., Żur, K.K., Kim, J. and Reddy, J.N., 2020. On the bifurcation buckling and vibration of porous nanobeams. *Composite Structures*, *250*, p. 112632.

Jena, S.K., Chakraverty, S., Mahesh, V. and Harursampath, D., 2022a. Application of Haar wavelet discretization and differential quadrature methods for free vibration of functionally graded micro-beam with porosity using modified couple stress theory. *Engineering Analysis with Boundary Elements*, *140*, pp. 167–185.

Jena, S.K., Chakraverty, S., Mahesh, V. and Harursampath, D., 2022b. Wavelet-based techniques for Hygro-Magneto-Thermo vibration of nonlocal strain gradient nanobeam resting on Winkler-Pasternak elastic foundation. *Engineering Analysis with Boundary Elements*, *140*, pp. 494–506.

Jena, S.K., Chakraverty, S., Mahesh, V., Harursampath, D. and Sedighi, H.M., 2022c. A novel numerical approach for the stability of nanobeam exposed to hygro-thermo-magnetic environment embedded in elastic foundation. *ZAMM Journal of Applied Mathematics and Mechanics/Zeitschrift für Angewandte Mathematik und Mechanik*, *102*(5), p. e202100380.

Jena, S.K., Chakraverty, S. and Malikan, M., 2021. Application of shifted Chebyshev polynomial-based Rayleigh—Ritz method and Navier's technique for vibration analysis of a functionally graded porous beam embedded in Kerr foundation. *Engineering with Computers*, *37*, pp. 3569–3589.

Jena, S.K., Chakraverty, S., Malikan, M. and Sedighi, H., 2020. Implementation of Hermite—Ritz method and Navier's technique for vibration of functionally graded porous nanobeam embedded in Winkler—Pasternak elastic foundation using bi-Helmholtz nonlocal elasticity. *Journal of Mechanics of Materials and Structures*, *15*(3), pp. 405–434.

Kanani, A.S., Niknam, H., Ohadi, A.R. and Aghdam, M.M., 2014. Effect of nonlinear elastic foundation on large amplitude free and forced vibration of functionally graded beam. *Composite Structures*, *115*, pp. 60–68.

Karamanli, A. and Vo, T.P., 2022. Finite element model for free vibration analysis of curved zigzag nanobeams. *Composite Structures*, *282*, p. 115097.

Ke, L.L. and Wang, Y.S., 2011. Size effect on dynamic stability of functionally graded microbeams based on a modified couple stress theory. *Composite Structures*, *93*(2), pp. 342–350.

Ke, L.L., Yang, J. and Kitipornchai, S., 2010. An analytical study on the nonlinear vibration of functionally graded beams. *Meccanica, 45*(6), pp. 743–752.

Khaniki, H.B., 2019. On vibrations of FG nanobeams. *International Journal of Engineering Science, 135*, pp. 23–36.

Khaniki, H.B. and Hosseini-Hashemi, S., 2017. Buckling analysis of tapered nanobeams using nonlocal strain gradient theory and a generalized differential quadrature method. *Materials Research Express, 4*(6), p. 065003.

Khastan, A., Nieto, J.J. and Rodriguez-Lopez, R., 2011. Variation of constant formula for first order fuzzy differential equations. *Fuzzy Sets and Systems, 177*(1), pp. 20–33.

Khastan, A., Nieto, J.J. and Rodriguez-Lopez, R., 2013. Periodic boundary value problems for first-order linear differential equations with uncertainty under generalized differentiability. *Information Sciences, 222*, pp. 544–558.

Koizumu, M., 1993. The concept of FGM, ceramic transactions. *Functionally Gradient Materials, 34*, pp. 3–10.

Koutsoumaris, C.C. and Eptaimeros, K., 2018. A research into bi-Helmholtz type of nonlocal elasticity and a direct approach to Eringen's nonlocal integral model in a finite body. *Acta Mechanica, 229*(9), pp. 3629–3649.

Krishnan, A., Dujardin, E., Ebbesen, T.W., Yianilos, P.N. and Treacy, M.M.J., 1998. Young's modulus of single-walled nanotubes. *Physical Review B, 58*(20), p. 14013.

Lam, D.C., Yang, F., Chong, A.C.M., Wang, J. and Tong, P., 2003. Experiments and theory in strain gradient elasticity. *Journal of the Mechanics and Physics of Solids, 51*(8), pp. 1477–1508.

Lazar, M., Maugin, G.A. and Aifantis, E.C., 2006. On a theory of nonlocal elasticity of bi-Helmholtz type and some applications. *International Journal of Solids and Structures, 43*(6), pp. 1404–1421.

Li, L., Hu, Y. and Li, X., 2016. Longitudinal vibration of size-dependent rods via nonlocal strain gradient theory. *International Journal of Mechanical Sciences, 115*, pp. 135–144.

Li, Z., Xu, Y. and Huang, D., 2021. Analytical solution for vibration of functionally graded beams with variable cross-sections resting on Pasternak elastic foundations. *International Journal of Mechanical Sciences, 191*, p. 106084.

Lim, C.W., Zhang, G. and Reddy, J., 2015. A higher-order nonlocal elasticity and strain gradient theory and its applications in wave propagation. *Journal of the Mechanics and Physics of Solids, 78*, pp. 298–313.

Liu, H. and Lv, Z., 2019. Vibration performance evaluation of smart magneto-electro-elastic nanobeam with consideration of nanomaterial uncertainties. *Journal of Intelligent Material Systems and Structures, 30*(18–19), pp. 2932–2952.

Liu, H., Lv, Z. and Li, Q., 2017. Flexural wave propagation in fluid-conveying carbon nanotubes with system uncertainties. *Microfluidics and Nanofluidics, 21*(8), pp. 1–13.

Lv, Z. and Liu, H., 2017. Nonlinear bending response of functionally graded nanobeams with material uncertainties. *International Journal of Mechanical Sciences, 134*, pp. 123–135.

Lv, Z. and Liu, H., 2018. Uncertainty modeling for vibration and buckling behaviors of functionally graded nanobeams in thermal environment. *Composite Structures, 184*, pp. 1165–1176.

Lv, Z., Liu, H. and Li, Q., 2018. Effect of uncertainty in material properties on wave propagation characteristics of nanorod embedded in elastic medium. *International Journal of Mechanics and Materials in Design, 14*(3), pp. 375–392.

Ma, W.L., Jiang, Z.C., Lee, K.Y. and Li, X.F., 2020. A refined beam theory for bending and vibration of functionally graded tube-beams. *Composite Structures, 236*, p. 111878.

Mahamood, R.M., Akinlabi, E.T., Shukla, M. and Pityana, S.L., 2012. Functionally graded material: An overview.

Malikan, M., Dimitri, R. and Tornabene, F., 2019. Transient response of oscillated carbon nanotubes with an internal and external damping. *Composites Part B: Engineering*, *158*, pp. 198–205.

McFarland, A.W. and Colton, J.S., 2005. Role of material microstructure in plate stiffness with relevance to micro-cantilever sensors. *Journal of Micromechanics and Microengineering*, *15*(5), p. 1060.

Mikaeilvand, N. and Khakrangin, S., 2012. Solving fuzzy partial differential equations by fuzzy two-dimensional differential transform method. *Neural Computing and Applications*, *21*(1), pp. 307–312.

Mindlin, R.D., 1962. *Influence of couple-stresses on stress concentrations*. Columbia Univ., New York.

Mindlin, R.D., 1965. Second gradient of strain and surface-tension in linear elasticity. *International Journal of Solids and Structures*, *1*(4), pp. 417–438.

Mindlin, R.D. and Tiersten, H., 1962. *Effects of couple-stresses in linear elasticity*. Columbia Univ., New York.

Mirjavadi, S.S., Afshari, B.M., Khezel, M., Shafiei, N., Rabby, S. and Kordnejad, M., 2018. Nonlinear vibration and buckling of functionally graded porous nanoscaled beams. *Journal of the Brazilian Society of Mechanical Sciences and Engineering*, *40*(7), pp. 1–12.

Muscolino, G. and Sofi, A., 2012. Stochastic analysis of structures with uncertain-but-bounded parameters via improved interval analysis. *Probabilistic Engineering Mechanics*, *28*, pp. 152–163.

Nateghi, A. and Salamat-talab, M., 2013. Thermal effect on size dependent behavior of functionally graded microbeams based on modified couple stress theory. *Composite Structures*, *96*, pp. 97–110.

Nazemnezhad, R. and Hosseini-Hashemi, S., 2014. Nonlocal nonlinear free vibration of functionally graded nanobeams. *Composite Structures*, *110*, pp. 192–199.

Park, S.K. and Gao, X.L., 2006. Bernoulli–Euler beam model based on a modified couple stress theory. *Journal of Micromechanics and Microengineering*, *16*(11), p. 2355.

Park, S.K. and Gao, X.L., 2008. Variational formulation of a modified couple stress theory and its application to a simple shear problem. *Zeitschrift für angewandte Mathematik und Physik*, *59*(5), pp. 904–917.

Peddieson, J., Buchanan, G.R. and McNitt, R.P., 2003. Application of nonlocal continuum models to nanotechnology. *International Journal of Engineering Science*, *41*(3–5), pp. 305–312.

Pei, J., Tian, F. and Thundat, T., 2004. Glucose biosensor based on the microcantilever. *Analytical Chemistry*, *76*(2), pp. 292–297.

Pradhan, K.K. and Chakraverty, S., 2013. Free vibration of Euler and Timoshenko functionally graded beams by Rayleigh—Ritz method. *Composites Part B: Engineering*, *51*, pp. 175–184.

Qiu, Z. and Wang, X., 2003. Comparison of dynamic response of structures with uncertain-but-bounded parameters using non-probabilistic interval analysis method and probabilistic approach. *International Journal of Solids and Structures*, *40*(20), pp. 5423–5439.

Rahimi, G.H., Gazor, M.S., Hemmatnezhad, M. and Toorani, H., 2013. On the postbuckling and free vibrations of FG Timoshenko beams. *Composite Structures*, *95*, pp. 247–253.

Rahimi, Z., Sumelka, W. and Yang, X.J., 2017. A new fractional nonlocal model and its application in free vibration of Timoshenko and Euler-Bernoulli beams. *The European Physical Journal Plus*, *132*(11), pp. 1–10.

Reddy, J., 2007. Nonlocal theories for bending, buckling and vibration of beams. *International Journal of Engineering Science, 45*(2–8), pp. 288–307.

Salvetat, J.P., Briggs, G.A.D., Bonard, J.M., Bacsa, R.R., Kulik, A.J., Stöckli, T., Burnham, N.A. and Forró, L., 1999. Elastic and shear moduli of single-walled carbon nanotube ropes. *Physical Review Letters, 82*(5), p. 944.

Selmi, A., 2020. Exact solution for nonlinear vibration of clamped-clamped functionally graded buckled beam. *Smart Structures and Systems, 26*(3), pp. 361–371.

Shafiei, N., Kazemi, M., Safi, M. and Ghadiri, M., 2016a. Nonlinear vibration of axially functionally graded nonuniform nanobeams. *International Journal of Engineering Science, 106*, pp. 77–94.

Shafiei, N., Mousavi, A. and Ghadiri, M., 2016b. Vibration behavior of a rotating nonuniform FG microbeam based on the modified couple stress theory and GDQEM. *Composite Structures, 149*, pp. 157–169.

Sharabiani, P.A. and Yazdi, M.R.H., 2013. Nonlinear free vibrations of functionally graded nanobeams with surface effects. *Composites Part B: Engineering, 45*(1), pp. 581–586.

Shooshtari, A. and Rafiee, M., 2011. Nonlinear forced vibration analysis of clamped functionally graded beams. *Acta Mechanica, 221*(1), pp. 23–38.

Şimşek, M., 2016. Nonlinear free vibration of a functionally graded nanobeam using nonlocal strain gradient theory and a novel Hamiltonian approach. *International Journal of Engineering Science, 105*, pp. 12–27.

Şimşek, M., 2019. Some closed-form solutions for static, buckling, free and forced vibration of functionally graded (FG) nanobeams using nonlocal strain gradient theory. *Composite Structures, 224*, p. 111041.

Şimşek, M. and Reddy, J.N., 2013. Bending and vibration of functionally graded microbeams using a new higher order beam theory and the modified couple stress theory. *International Journal of Engineering Science, 64*, pp. 37–53.

Sina, S.A., Navazi, H.M. and Haddadpour, H., 2009. An analytical method for free vibration analysis of functionally graded beams. *Materials & Design, 30*(3), pp. 741–747.

Sınır, S., Çevik, M. and Sınır, B.G., 2018. Nonlinear free and forced vibration analyses of axially functionally graded Euler-Bernoulli beams with nonuniform cross-section. *Composites Part B: Engineering, 148*, pp. 123–131.

Su, H. and Banerjee, J.R., 2015. Development of dynamic stiffness method for free vibration of functionally graded Timoshenko beams. *Computers & Structures, 147*, pp. 107–116.

Tapaswini, S. and Chakraverty, S., 2014. Dynamic response of imprecisely defined beam subject to various loads using Adomian decomposition method. *Applied Soft Computing, 24*, pp. 249–263.

Thai, H.T. and Vo, T.P., 2012. Bending and free vibration of functionally graded beams using various higher-order shear deformation beam theories. *International Journal of Mechanical Sciences, 62*(1), pp. 57–66.

Tossapanon, P. and Wattanasakulpong, N., 2016. Stability and free vibration of functionally graded sandwich beams resting on two-parameter elastic foundation. *Composite Structures, 142*, pp. 215–225.

Toupin, R., 1962. Elastic materials with couple-stresses. *Archive for Rational Mechanics and Analysis, 11*(1), pp. 385–414.

Uzun, B. and Yaylı, M.Ö., 2019. Finite element model of functionally graded nanobeam for free vibration analysis. *International Journal of Engineering and Applied Sciences, 11*(2), pp. 387–400.

Uzun, B. and Yaylı, M.Ö., 2020. Nonlocal vibration analysis of Ti-6Al-4V/ZrO 2 functionally graded nanobeam on elastic matrix. *Arabian Journal of Geosciences, 13*(4), pp. 1–10.

Vo, T.P., Thai, H.T., Nguyen, T.K., Maheri, A. and Lee, J., 2014. Finite element model for vibration and buckling of functionally graded sandwich beams based on a refined shear deformation theory. *Engineering Structures*, *64*, pp. 12–22.

Wang, C.M., Zhang, Y.Y. and He, X.Q., 2007. Vibration of nonlocal Timoshenko beams. *Nanotechnology*, *18*(10), p. 105401.

Wang, C.M., Zhang, Y.Y., Ramesh, S.S. and Kitipornchai, S., 2006. Buckling analysis of micro-and nano-rods/tubes based on nonlocal Timoshenko beam theory. *Journal of Physics D: Applied Physics*, *39*(17), p. 3904.

Wang, S.S., 1983. Fracture mechanics for delamination problems in composite materials. *Journal of Composite Materials*, *17*(3), pp. 210–223.

Wattanasakulpong, N., Prusty, B.G. and Kelly, D.W., 2011. Thermal buckling and elastic vibration of third-order shear deformable functionally graded beams. *International Journal of Mechanical Sciences*, *53*(9), pp. 734–743.

Wattanasakulpong, N., Prusty, B.G., Kelly, D.W. and Hoffman, M., 2012. Free vibration analysis of layered functionally graded beams with experimental validation. *Materials & Design (1980–2015)*, *36*, pp. 182–190.

Xu, M., 2006. Free transverse vibrations of nano-to-micron scale beams. *Proceedings of the Royal Society A: Mathematical, Physical and Engineering Sciences*, *462*(2074), pp. 2977–2995.

Xu, Y., Qian, Y., Chen, J. and Song, G., 2015. Stochastic dynamic characteristics of FGM beams with random material properties. *Composite Structures*, *133*, pp. 585–594.

Yang, F.A.C.M., Chong, A.C.M., Lam, D.C.C. and Tong, P., 2002. Couple stress based strain gradient theory for elasticity. *International Journal of Solids and Structures*, *39*(10), pp. 2731–2743.

Yang, J.F.C. and Lakes, R.S., 1982. Experimental study of micropolar and couple stress elasticity in compact bone in bending. *Journal of Biomechanics*, *15*(2), pp. 91–98.

Yu, Y.J., Xue, Z.N., Li, C.L. and Tian, X.G., 2016. Buckling of nanobeams under nonuniform temperature based on nonlocal thermoelasticity. *Composite Structures*, *146*, pp. 108–113.

Yu, Y.J., Zhang, K. and Deng, Z.C., 2019. Buckling analyses of three characteristic-lengths featured size-dependent gradient-beam with variational consistent higher order boundary conditions. *Applied Mathematical Modelling*, *74*, pp. 1–20.

Zadeh, L.A., 1996. Fuzzy sets. In *Fuzzy sets, fuzzy logic, and fuzzy systems: Selected papers by Lotfi A Zadeh* (pp. 394–432).

Zeighampour, H. and Beni, Y.T., 2015. Free vibration analysis of axially functionally graded nanobeam with radius varies along the length based on strain gradient theory. *Applied Mathematical Modelling*, *39*(18), pp. 5354–5369.

2 Preliminaries

2.1 INTRODUCTION

Fuzzy set theory serves as a powerful mathematical framework for handling uncertainty and imprecision in various real-world applications. This chapter provides a comprehensive overview of fundamental concepts in fuzzy set theory, including definitions of fuzzy sets and different types of fuzzy numbers such as triangular, trapezoidal, and Gaussian fuzzy numbers. Additionally, the concept of double parametric form, which plays a crucial role in representing uncertain parameters, is introduced. To delve deeper into the nuances of fuzzy set theory, renowned works by Zimmermann [1], Jaulin et al. [2], Ross [3], Hanss [4], Moore [5], Chakraverty [6], and Chakraverty et al. [7–9] are recommended, providing readers with valuable insights into advanced topics in fuzzy set theory. This chapter serves as a concise introduction, laying the foundation for the subsequent chapters. It draws inspiration from the other works of same authors, Chakraverty et al. [10], providing a framework for the topics explored further in this book.

Prior to delineating the fuzzy set, it is imperative to introduce the fundamental concepts of intervals.

2.2 INTERVAL

Intervals are fundamental in various areas of mathematics, including calculus, analysis, and set theory. They are commonly used to represent ranges of values, such as time intervals, temperature ranges, or confidence intervals in statistics. Interval notation provides a concise and precise way to describe sets of real numbers, making it a valuable tool in mathematical reasoning and problem-solving. In mathematics, an interval is a set of real numbers that includes all the numbers between any two given numbers within the set. It is defined by its endpoints, which are usually denoted by a lower bound and an upper bound.

Let's consider an interval $\tilde{I}$, dented by $[\underline{I}, \overline{I}]$, on the set of real number $\mathbb{R}$ as defined by:

$$\tilde{I} = [\underline{I}, \overline{I}] = \left\{ I \in \mathbb{R} : \underline{I} \leq I \leq \overline{I} \right\} \tag{2.1}$$

where $\underline{I}$ and $\overline{I}$ represent lower bound and upper bound or the left and right endpoints of the interval. Intervals can take various forms, including closed, open, half-open, or half-closed, depending on whether their endpoints are inclusive or exclusive. However, for consistency, this book exclusively considers closed intervals.

DOI: 10.1201/9781003303107-2

The two arbitrary intervals, denoted as $\tilde{I} = [\underline{I}, \overline{I}]$ and $\tilde{\Psi} = [\underline{\Psi}, \overline{\Psi}]$, are considered equal if they belong to the same set, which can be mathematically expressed as:

$$\tilde{I} = \tilde{\Psi} \text{ if and only if } \underline{I} = \underline{\Psi} \text{ and } \overline{I} = \overline{\Psi} \qquad (2.2)$$

2.2.1 INTERVAL ARITHMETIC

Interval arithmetic is a mathematical technique used to perform calculations involving intervals, which are sets of real numbers bounded by two endpoints. In the context of fuzzy theory, interval arithmetic is extended to handle fuzzy intervals, where each endpoint of the interval is itself a fuzzy number representing uncertainty or imprecision. Fuzzy intervals allow for the representation of uncertain quantities with varying degrees of fuzziness, enabling more realistic modeling of uncertain systems. Interval arithmetic with fuzzy intervals involves performing arithmetic operations (addition, subtraction, multiplication, and division) on fuzzy numbers or fuzzy intervals. The result of these operations is itself a fuzzy interval, reflecting the uncertainty inherent in the input fuzzy intervals.

The interval arithmetic operations of addition ($+$), subtraction ($-$), multiplication ($\times$), and division ($\div$) for the given two arbitrary intervals $\tilde{I} = [\underline{I}, \overline{I}]$ and $\tilde{\Psi} = [\underline{\Psi}, \overline{\Psi}]$ are defined as follows:

(a) $\tilde{I} + \tilde{\Psi} = [\underline{I} + \underline{\Psi}, \overline{I} + \overline{\Psi}]$

(b) $\tilde{I} - \tilde{\Psi} = [\underline{I} - \overline{\Psi}, \overline{I} - \underline{\Psi}]$

(c) $\tilde{I} \times \tilde{\Psi} = \left[\min\left(\underline{I} \times \underline{\Psi}, \underline{I} \times \overline{\Psi}, \overline{I} \times \underline{\Psi}, \overline{I} \times \overline{\Psi}\right), \max\left(\underline{I} \times \underline{\Psi}, \underline{I} \times \overline{\Psi}, \overline{I} \times \underline{\Psi}, \overline{I} \times \overline{\Psi}\right)\right]$

(d) $\dfrac{\tilde{I}}{\tilde{\Psi}} = [\underline{I}, \overline{I}] \times \left[\dfrac{1}{\overline{\Psi}}, \dfrac{1}{\underline{\Psi}}\right]$, if $0 \notin \tilde{\xi}$

(e) $k\tilde{I} = \begin{cases} \left[k\overline{I}, k\underline{I}\right], & k < 0, \\ \left[k\underline{I}, k\overline{I}\right], & k \geq 0. \end{cases}$

where k is a real number.

2.3 FUZZY SET

A fuzzy set is a mathematical representation used to model uncertainty or vagueness in data. Unlike classical sets, where an element is either a member or not, a fuzzy set allows for partial membership. In a fuzzy set, each element is assigned a degree of membership ranging between 0 and 1, indicating the degree to which the element belongs to the set. The degree of membership represents the extent to which an element possesses the characteristics defined by the set. Formally, let X be a universe of discourse or a set of objects, and let $\mu_A(x)$ denote the membership function of a fuzzy set A defined on X. The membership function assigns a degree of membership, denoted by $\mu_A(x)$, to each element x in X, indicating the extent to which x belongs to the fuzzy set A. The definition of a fuzzy set A in X is given by:

$$\tilde{A} = \left\{ (x, \mu_A(x)) : x \in X, \mu_A(x) \in [0,1] \right\} \tag{2.3}$$

where $(x, \mu_A(x))$ represents an ordered pair consisting of an element x from X and its corresponding membership degree $\mu_A(x)$ in the fuzzy set A, and it is piecewise continuous. This definition allows for the representation of vague or imprecise information, making fuzzy sets valuable in various fields, including artificial intelligence, control systems, pattern recognition, and decision-making.

2.4 FUZZY NUMBER

A fuzzy number is a mathematical construct used to represent uncertain or imprecise numerical values. Unlike crisp numbers, which have precise values, fuzzy numbers allow for the modeling of uncertainty by assigning a range of possible values along with a degree of membership indicating the likelihood of each value within that range.

Mathematically, a fuzzy number $\tilde{A}$ is a convex normalized fuzzy set $\tilde{A}$ of the real line $\mathbb{R}$ such that

$$\left\{ \mu_{\tilde{A}}(x) : \mathbb{R} \rightarrow [0,1], \forall x \in \mathbb{R} \right\} \tag{2.4}$$

where $\mu_{\tilde{A}}$ is a membership function, and it is piecewise continuous. Several types of fuzzy numbers are recognized in literature; yet, this discussion focuses solely on three main types: Triangular, trapezoidal, and Gaussian fuzzy numbers.

2.4.1 TRIANGULAR FUZZY NUMBER

A Triangular Fuzzy Number is characterized by a triangular-shaped membership function, as illustrated in Figure 2.1, which assigns a degree of membership to each value within a specified range. It is defined by three parameters: The lower bound (a_1), the upper bound (c_1), and the modal value (b_1), where $a_1 \leq b_1 \leq c_1$. The membership function is typically triangular in shape, with the degree of membership being highest at the modal value and decreasing linearly towards the lower and upper bounds.

A Triangular Fuzzy Number $\tilde{A}$ is a convex normalized fuzzy set $\tilde{A}$ of the real line $\mathbb{R}$ such that:

(a) There exists exactly one $x_0 \in \mathbb{R}$ such that $\mu_{\tilde{A}}(x_0) = 1$, where x_0 is called the mean value of $\tilde{A}$, and $\mu_{\tilde{A}}$ is called the membership function of the fuzzy set.

(b) $\mu_{\tilde{A}}(x)$ is piecewise continuous.

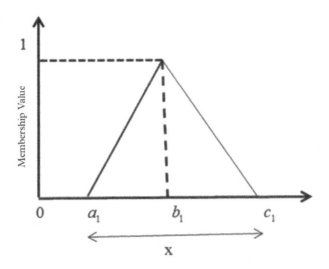

FIGURE 2.1   Schematic representation of Triangular Fuzzy Numbers.

Let us consider a Triangular Fuzzy Number, $\tilde{A} = (a_1, b_1, c_1)$. Then, the membership function $\mu_{\tilde{A}}$ of the Triangular Fuzzy Number $\tilde{A} = (a_1, b_1, c_1)$ is defined as follows:

$$\mu_{\tilde{A}}(x) = \begin{cases} 0, & x \le a_1, \\ \dfrac{x - a_1}{b_1 - a_1}, & a_1 \le x \le b_1, \\ \dfrac{c_1 - x}{c_1 - b_1}, & b_1 \le x \le c_1, \\ 0, & x \ge c_1. \end{cases} \tag{2.5}$$

2.4.1.1   Single Parametric Form of Triangular Fuzzy Number

The Triangular Fuzzy Number $\tilde{A} = (a_1, b_1, c_1)$ can be represented by an ordered pair of functions through $\alpha - cut$ approach as:

$$\tilde{A} = (a_1, b_1, c_1) = \left[\underline{A}(\alpha), \overline{A}(\alpha) \right] = \left[(b_1 - a_1)\alpha + a_1, \ -(c_1 - b_1)\alpha + c_1 \right],$$

$$\text{with } \alpha \in [0, 1] \tag{2.6}$$

2.4.1.2   Double Parametric Form of Triangular Fuzzy Number

Applying α-cut approach, the Triangular Fuzzy Number can be converted into the interval form as $\tilde{A} = \left[\underline{A}(\alpha), \overline{A}(\alpha) \right]$. Now, by using another parameter β, the interval form can be written in crisp form as:

$$\tilde{A}(\alpha, \beta) = \left[\underline{A}(\alpha), \overline{A}(\alpha) \right] = \beta \left(\overline{A}(\alpha) - \underline{A}(\alpha) \right) + \underline{A}(\alpha), \text{ with } \alpha, \beta \in [0, 1] \tag{2.7}$$

Lower and upper bound solution in single parametric form can be obtained by substituting $\beta = 0$ and $\beta = 1$, respectively. Mathematically, these can be represented as $\tilde{A}(\alpha, 0) = \underline{A}(\alpha)$ and $\tilde{A}(\alpha, 1) = \overline{A}(\alpha)$.

2.4.2 TRAPEZOIDAL FUZZY NUMBER

A trapezoidal fuzzy number is a fuzzy number characterized by a trapezoidal membership function, as illustrated in Figure 2.2. It is defined by four parameters: (a_1, b_1, c_1, d_1), where a_1 and d_1 represent the lower and upper bounds of the support, respectively, and b_1 and c_1 are the two interior points where the membership function reaches its maximum value of 1. Let's consider an arbitrary trapezoidal fuzzy number $\tilde{A} = (a_1, b_1, c_1, d_1)$ with membership function $\mu_{\tilde{A}}$ of $\tilde{A} = (a_1, b_1, c_1, d_1)$ written as:

$$\mu_{\tilde{A}}(x) = \begin{cases} 0, & x \leq a_1, \\ \dfrac{x - a_1}{b_1 - a_1}, & a_1 \leq x \leq b_1, \\ 1, & b_1 \leq x \leq c_1, \\ \dfrac{d_1 - x}{d_1 - c_1}, & c_1 \leq x \leq d_1, \\ 0, & x \geq d_1. \end{cases} \tag{2.8}$$

2.4.2.1 Single Parametric Form of Trapezoidal Fuzzy Number

The trapezoidal fuzzy number $\tilde{A} = (a_1, b_1, c_1, d_1)$ can be represented by an ordered pair of functions through $\alpha - cut$ approach as:

$$\tilde{A} = (a_1, b_1, c_1, d_1) = \left[\underline{A}(\alpha), \overline{A}(\alpha)\right] = \left[(b_1 - a_1)\alpha + a_1, -(d_1 - c_1)\alpha + d_1\right],$$
$$\text{with } \alpha \in [0, 1] \tag{2.9}$$

2.4.2.2 Double Parametric Form of Trapezoidal Fuzzy Number

Applying α-cut approach, the trapezoidal fuzzy number can be converted into the interval form as $\tilde{A} = \left[\underline{A}(\alpha), \overline{A}(\alpha)\right]$. Now, by using another parameter β, the interval form can be written in crisp form as:

$$\tilde{A}(\alpha, \beta) = \left[\underline{A}(\alpha), \overline{A}(\alpha)\right] = \beta\left(\overline{A}(\alpha) - \underline{A}(\alpha)\right) + \underline{A}(\alpha), \text{ with } \alpha, \beta \in [0, 1] \tag{2.10}$$

Lower and upper bound solution in a single parametric form can be obtained by substituting $\beta = 0$ and $\beta = 1$, respectively. Mathematically, these can be represented as $\tilde{A}(\alpha, 0) = \underline{A}(\alpha)$ and $\tilde{A}(\alpha, 1) = \overline{A}(\alpha)$.

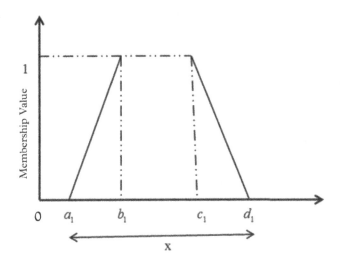

FIGURE 2.2 Schematic representation of trapezoidal fuzzy numbers.

2.4.3 GAUSSIAN FUZZY NUMBER

A Gaussian fuzzy number is a type of fuzzy number characterized by a Gaussian or bell-shaped membership function, as illustrated in Figure 2.3. It is defined by two parameters: The mean and the standard deviation. Let us consider an arbitrary standard Gaussian fuzzy number, $\tilde{A} = (\eta, \lambda, \lambda)$, with the corresponding membership function written as:

$$\mu_{\tilde{A}}(x) = \exp\left(\frac{-(x-\eta)^2}{2\lambda^2}\right), \forall x \in \mathbb{R} \text{ and where } \beta = \frac{1}{2\lambda^2} \qquad (2.11)$$

where η represents the mean or modal value and λ denotes the standard deviation or spread of fuzziness.

2.4.3.1 Single Parametric Form of Gaussian Fuzzy Number

The symmetric or standard Gaussian fuzzy number $\tilde{A} = (\eta, \lambda, \lambda)$ can be expressed in single parametric form by an ordered pair of functions through $\alpha - cut$ approach as:

$$\tilde{A} = (\eta, \lambda, \lambda) = \left[\underline{A}(\alpha), \overline{A}(\alpha)\right] = \left[\eta - \sqrt{-2\lambda^2(\log_e \alpha)}, \eta + \sqrt{-2\lambda^2(\log_e \alpha)}\right]$$

$$\text{with } \alpha \in (0, 1] \qquad (2.12)$$

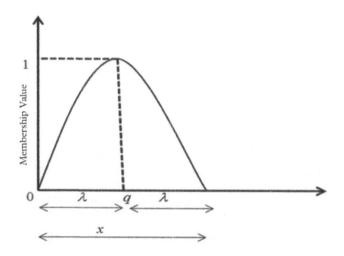

FIGURE 2.3 Schematic representation of Gaussian fuzzy numbers.

2.4.3.2 Double Parametric Form of Gaussian Fuzzy Number

Applying α-cut approach, symmetric or standard Gaussian fuzzy number can be converted into interval form as $\tilde{A} = \left[\underline{A}(\alpha),\, \overline{A}(\alpha)\right]$. Now, by using another parameter β, the interval form can be written in crisp form as:

$$\tilde{A}(\alpha, \beta) = \left[\underline{A}(\alpha),\, \overline{A}(\alpha)\right] = \beta\left(\overline{A}(\alpha) - \underline{A}(\alpha)\right) + \underline{A}(\alpha), \text{ with } \alpha \in (0,\, 1]$$
$$\text{and } \beta \in [0,\, 1] \tag{2.13}$$

Lower and upper bound solution in single parametric form can be obtained by substituting $\beta = 0$ and $\beta = 1$, respectively. Mathematically, these can be represented as $\tilde{A}(\alpha, 0) = \underline{A}(\alpha)$ and $\tilde{A}(\alpha, 1) = \overline{A}(\alpha)$.

For all the above-mentioned fuzzy numbers, the lower and upper bounds of the fuzzy numbers satisfy the following:

(i) $\underline{A}(\alpha)$ is a bounded left-continuous nondecreasing function over $[0,\, 1]$.
(ii) $\overline{A}(\alpha)$ is a bounded right-continuous nonincreasing function over $[0,1]$.
(iii) $\underline{A}(\alpha) \leq \overline{A}(\alpha)$, where $\alpha \in [0,\, 1]$.

2.5 FUZZY ARITHMETIC

For the given two arbitrary fuzzy numbers $\tilde{A} = \left[\underline{A}(\alpha),\, \overline{A}(\alpha)\right]$ and $\tilde{B} = \left[\underline{B}(\alpha),\, \overline{B}(\alpha)\right]$, the fuzzy arithmetic is similar to the interval arithmetic which is defined as:

(a) $\tilde{A} + \tilde{B} = \left[\underline{A}(\alpha) + \underline{B}(\alpha),\, \overline{A}(\alpha) + \overline{B}(\alpha)\right]$
(b) $\tilde{A} - \tilde{B} = \left[\underline{A}(\alpha) - \overline{B}(\alpha),\, \overline{A}(\alpha) - \underline{B}(\alpha)\right]$

(c) $\bar{A} \times \bar{B} = \begin{bmatrix} \min\left(\underline{A}(\alpha) \times \underline{B}(\alpha), \underline{A}(\alpha) \times \bar{B}(\alpha), \bar{A}(\alpha) \times \underline{B}(\alpha), \bar{A}(\alpha) \times \bar{B}(\alpha)\right), \\ \max\left(\underline{A}(\alpha) \times \underline{B}(\alpha), \underline{A}(\alpha) \times \bar{B}(\alpha), \bar{A}(\alpha) \times \underline{B}(\alpha), \bar{A}(\alpha) \times \bar{B}(\alpha)\right) \end{bmatrix}$

(d) $\dfrac{\tilde{A}}{\tilde{B}} = \left[\underline{A}(\alpha), \bar{A}(\alpha)\right] \times \left[\dfrac{1}{\bar{B}(\alpha)}, \dfrac{1}{\underline{B}(\alpha)}\right]$ if $0 \notin \tilde{B}(\alpha)$

(e) $k\tilde{A} = \begin{cases} \left[k\bar{A}(\alpha), k\underline{A}(\alpha)\right], & k < 0, \\ \left[k\underline{A}(\alpha), k\bar{A}(\alpha)\right], & k \geq 0. \end{cases}$ where k is a real number.

BIBLIOGRAPHY

[1] Zimmermann, H. J., Fuzzy Set Theory—and Its Applications, Dordrecht, Springer Science & Business Media, 2011.
[2] Jaulin, L., Kieffer, M., Didri, O. T., Walterm, E., Applied Interval Analysis, London, Springer, 2001.
[3] Ross, T. J., Fuzzy Logic with Engineering Applications, New York, John Wiley & Sons, 2004.
[4] Hanss, M., Applied Fuzzy Arithmetic: An Introduction with Engineering Applications, Berlin, Springer-Verlag, 2005.
[5] Moore, R. E., Interval Analysis, Englewood Cliffs, Prentice-Hall, 1966.
[6] Chakraverty, S., Mathematics of Uncertainty Modeling in the Analysis of Engineering and Science Problems, Hershey, IGI Global Publication, 2014.
[7] Chakraverty, S., Tapaswini, S., Behera, D., Fuzzy Arbitrary Order System: Fuzzy Fractional Differential Equations and Applications, Hoboken, NJ, John Wiley & Sons, 2016.
[8] Chakraverty, S., Tapaswini, S., Behera, D., Fuzzy Differential Equations and Applications for Engineers and Scientists, Boca Raton, Taylor and Francis Group; CRC Press, 2016.
[9] Chakraverty, S., Sahoo, D. M., Mahato, N. R., Concepts of Soft Computing: Fuzzy and ANN with Programming, Singapore, Springer, 2019.
[10] Chakraverty, S., Jena, R. M. and Jena, S. K., Time- Fractional Order Biological Systems with Uncertain Parameters, Switzerland, Springer Nature, 2022.

3 Vibration of Euler–Bernoulli Nanobeam with Uncertainty

3.1 INTRODUCTION

This chapter delves into the modeling of an Euler–Bernoulli nanobeam within the framework of Eringen's nonlocal theory, incorporating material uncertainties associated with mass density and Young's modulus represented by fuzzy numbers, specifically Triangular Fuzzy Numbers. A novel double parametric Rayleigh–Ritz method is introduced to effectively address these uncertainties, enabling the investigation of vibration characteristics and the analysis of uncertainty propagation in frequency parameters. Material uncertainties are explored across three cases: (i) Imprecisely defined Young's modulus, (ii) imprecisely defined mass density, and (iii) simultaneous imprecision in both Young's modulus and mass density. Frequency parameters and mode shapes are computed and presented for both pinned–pinned (P-P) and clamped–clamped (C-C) boundary conditions. To validate the accuracy and efficiency of the models, convergence studies are conducted for all three cases. Lower and upper bounds of frequency parameters are computed using the double parameter approach, and graphical representations are generated as Triangular Fuzzy Numbers to illustrate the models' sensitivity to uncertainties. Comparative analyses with existing literature demonstrate robust agreement in special cases, highlighting the reliability of the proposed methodology. The uncertainty modeling and frequency parameter bounds established herein offer valuable insights for the design and optimization of engineering structures, facilitating enhanced quality and performance.

3.2 FORMULATION OF THE PROPOSED MODEL

The displacement fields of the proposed model are based on the Euler–Bernoulli beam theory. To describe the kinematics of the model, the x axis is taken along the length of the beam whereas y axis and z axis are considered along the width and thickness of the beam, respectively. Here, all the displacement fields (u_1, u_2, u_3) are the displacements along x, y, and z coordinates, respectively. The displacement fields of Euler–Bernoulli beam theory is given as [1]:

$$
\begin{aligned}
u_1(x,z) &= -z\frac{\partial w(x,t)}{\partial x}\\
u_2(x,z) &= 0\\
u_3(x,z) &= w(x,t)
\end{aligned}
\tag{3.1}
$$

DOI: 10.1201/9781003303107-3

where w is the transverse displacement of the point $(x,0)$ on the mid-plane $(z=0)$ and t denotes time. Here, the displacement along y coordinate u_2 is taken to be zero.

The strain energy U of the Euler–Bernoulli beam is expressed as:

$$U = \frac{1}{2}\int_0^L \int_A \sigma_{xx}\varepsilon_{xx}\, dA\, dx, \tag{3.2}$$

where σ_{xx} is the normal stress, L is the length of the beam, and A is the cross-section area. The strain–displacement relation may be given as:

$$\varepsilon_{xx} = -z\frac{\partial^2 w}{\partial x^2}, \tag{3.3}$$

where ε_{xx} is the normal strain and w is the deflection function. Substituting Eq. (3.3) in Eq. (3.2), one can obtain:

$$U = -\frac{1}{2}\int_0^L M\frac{\partial^2 w}{\partial x^2}\, dx, \tag{3.4}$$

where $M = \int_A z\sigma_{xx}\, dA$ is the bending moment. Here, we have assumed the transverse displacement as sinusoidal, i.e., $w(x,t) = w_0(x)e^{i\omega t}$ where ω is the natural frequency of vibration. The strain energy U is expressed as:

$$U = -\frac{1}{2}\int_0^L M\frac{d^2 w_0}{dx^2}\, dx. \tag{3.5}$$

The kinetic energy T is also given as

$$T = \frac{1}{2}\int_0^L \rho A\omega^2 w_0^2\, dx, \tag{3.6}$$

where ρ is the mass density and A is the area of beams. Using Hamilton's principle $\int_0^t (\delta T - \delta U)\, dt = 0$ and setting the coefficient of δw_0 to zero, one may obtain the governing equation as:

$$\frac{d^2 M}{dx^2} = -\rho A\omega^2 w_0 \tag{3.7}$$

For one-dimensional elastic material, nonlocal elasticity theory of Eringen [2] is given as:

$$\left(1 - \mu\frac{\partial^2}{\partial x^2}\right)\sigma_{xx} = E\varepsilon_{xx} \tag{3.8}$$

where $\mu = (e_0 a)^2$ is the nonlocal parameter, E is Young's modulus. Here, e_0 and a denote material constant and internal characteristic length, respectively.

Multiplying Eq. (3.8) by zdA and integrating over A, the nonlocal constitutive relation for Euler–Bernoulli nanobeam may be expressed as:

$$M - \mu \frac{d^2 M}{dx^2} = -EI \frac{d^2 w_0}{dx^2} \tag{3.9}$$

where $I = \int_A z^2 dA$ is the second moment of area. Now by plugging Eq. (3.9) and Eq. (3.7) in Eq. (3.5), we have:

$$U = -\frac{1}{2} \int_0^L \left(-EI \frac{d^2 w_0}{dx^2} - \mu \rho A \omega^2 w_0 \right) \frac{d^2 w_0}{dx^2} dx. \tag{3.10}$$

3.2.1 MODELING WITH MATERIAL UNCERTAINTIES

Material uncertainties in Young's modulus and mass density have been taken as:

$$\tilde{E} = E E_0 \quad \text{and} \quad \tilde{\rho} = \rho \rho_0 \tag{3.11}$$

where $\tilde{E} = (e_1, e_2, e_3)$ and $\tilde{\rho} = (\rho_1, \rho_2, \rho_3)$ are Triangular Fuzzy Numbers.

Using Eq. (3.11) in Eqs. (3.6) and (3.10), strain energy $(\tilde{U})$ and kinetic energy $(\tilde{T})$ of the uncertain system are obtained as:

$$\tilde{U} = \frac{1}{2} \int_0^L \left(\tilde{E} E_0 I \frac{d^2 w_0}{dx^2} + \mu \tilde{\rho} \rho_0 A \omega^2 w_0 \right) \frac{d^2 w_0}{dx^2} dx \tag{3.12}$$

$$\tilde{T} = \frac{1}{2} \int_0^L \tilde{\rho} \rho_0 A \omega^2 w_0^2 dx \tag{3.13}$$

Now, by equating the above uncertain strain and kinetic energies of the nanobeam, one may have:

$$\frac{1}{2} \int_0^L \left(\tilde{E} E_0 I \frac{d^2 w_0}{dx^2} + \mu \tilde{\rho} \rho_0 A \omega^2 w_0 \right) \frac{d^2 w_0}{dx^2} dx = \frac{1}{2} \int_0^L \tilde{\rho} \rho_0 A \omega^2 w_0^2 dx \tag{3.14}$$

Let us now introduce the following nondimensional terms:

$$X = \frac{x}{L} = \text{Nondimensional coordinate}$$

$$W = \frac{w_0}{L} = \text{Nondimensional transverse displacement}$$

$$\lambda^2 = \frac{\rho_0 A \omega^2 L^4}{EI_0} = \text{Nondimensional frequency parameter}$$

$$\tau = \frac{e_0 a}{L} = \text{Nondimensional nonlocal parameter}$$

Substituting the above nondimensional terms, Eq. (3.14) is reduced to:

$$\int_0^L \left(\frac{d^2 W}{dX^2}\right)^2 dX = \frac{\tilde{\rho}}{\tilde{E}} \lambda^2 \left[\int_0^1 W^2 dX - \tau^2 \int_0^1 W \frac{d^2 W}{dX^2} dX\right] \qquad (3.15)$$

Implementing the concept of double parametric form (as explained in preliminaries), the Triangular Fuzzy Numbers $\tilde{E} = (e_1, e_2, e_3)$ and $\tilde{\rho} = (\rho_1, \rho_2, \rho_3)$ can be written in double parametric form as:

$$\tilde{E}(\alpha, \beta) = (e_1 - e_3)\alpha\beta + (e_3 - e_1)\beta + (e_2 - e_1)\alpha + e_1 \quad \alpha, \beta \in [0\ 1] \qquad (3.16)$$

$$\tilde{\rho}(\alpha, \beta) = (\rho_1 - \rho_3)\alpha\beta + (\rho_3 - \rho_1)\beta + (\rho_2 - \rho_1)\alpha + \rho_1 \quad \alpha, \beta \in [0\ 1] \qquad (3.17)$$

Plugging Eq. (3.16) and Eq. (3.17) in Eq. (3.15), we get:

$$\int_0^L \left(\frac{d^2 W}{dX^2}\right)^2 dX = \frac{(e_1 - e_3)\alpha\beta + (e_3 - e_1)\beta + (e_2 - e_1)\alpha + e_1}{(\rho_1 - \rho_3)\alpha\beta + (\rho_3 - \rho_1)\beta + (\rho_2 - \rho_1)\alpha + \rho_1} \lambda^2$$
$$\left[\int_0^1 W^2 dX - \tau^2 \int_0^1 W \frac{d^2 W}{dX^2} dX\right] \qquad (3.18)$$

Now, Eq. (3.18) is in double parametric form for the uncertain system where α and β control the uncertainties. Finally, Eq. (3.18) may be solved for frequency parameters having the essence of various uncertainties with respect to different values of α and β. Next, we summarize below the Rayleigh–Ritz method where the double parametric forms of the properties may be introduced with respect to the parameters α and β.

3.3   SOLUTION METHODOLOGY OF THE PROPOSED PROBLEM

Choosing simple polynomials as basis functions in the Rayleigh–Ritz method, frequency parameters of the imprecisely defined system have been computed. In this approach, displacement function $W(X)$ is approximated by a series of admissible functions as:

$$W(X) = \sum_{i=1}^k a_i \theta_i = a_1 \theta_1 + a_2 \theta_2 + \cdots + a_{k-1} \theta_{k-1} + a_k \theta_k \qquad (3.19)$$

where $a_i's$ are unknown coefficients and $\theta_i's$ are admissible functions which are defined as:

$$\theta_i = \xi_b X^{i-1}, \quad i = 1,2,3, \cdots k \tag{3.20}$$

where ξ_b is the dimensionless boundary polynomials that regulate the boundary conditions and is defined as $\xi_b = X^m (1-X)^n$. Here, m and n take the values like 0, 1, 2 for free, simply supported, and clamped boundary conditions, respectively. Now substituting Eq. (3.19) in Eq. (3.18) and differentiating partially with respect to a_i, Eq. (18) may be reduced to a generalized eigenvalue problem as:

$$[S]\{A\} = \lambda^2 [T]\{A\} \tag{3.21}$$

where $\{A\} = [a_1, a_2, \cdots a_k]^T$, S is the stiffness matrix, and T is the mass matrix. It is again worth mentioning here that all the matrices as above will contain the two parameters α and β which control the uncertainties for different scenarios.

3.4 NUMERICAL RESULTS AND DISCUSSION

The customized codes in MATLAB interface have been developed to solve the generalized eigenvalue problem obtained in Eq. (3.21); and frequency parameters $\sqrt{\lambda}$ have been computed for all the three cases such as: (i) Young's modulus is imprecisely defined, (ii) mass density is imprecisely defined, and (iii) both Young's modulus and mass density are imprecisely defined. We go through all the cases one by one for pinned–pinned (P-P) and clamped–clamped (C-C) edges. For the computational purpose, we have considered the parameters as $E = \tilde{E}E_0$, $\tilde{E} = (0.5, 1, 1.5)$, $E_0 = 30 \times 10^6$, $\rho = \tilde{\rho}\rho_0$, $\tilde{\rho} = (0.8, 1, 1.2)$, $L = 10\,nm$, and $\dfrac{L}{h} = 10$.

3.4.1 VALIDATION OF THE PROPOSED MODEL

For the validation of the frequency parameters, $\alpha = 1$ and $\beta = 0$ have been set in the proposed model. Moreover, the obtained results from the present model using the Rayleigh–Ritz method are compared with that from References [3–4] for pinned–pined (P-P) and clamped–clamped (C-C) boundary conditions, which are demonstrated in Tables 3.1–3.4. From the Tables 3.1–3.4, one may witness a robust

---

Table 3.1

Validation of the Present Result with [3] for P-P Case

$\tau = \dfrac{e_0 a}{L}$	First Mode		Second Mode		Third Mode		Fourth Mode	
	Present	[3]	Present	[3]	Present	[3]	Present	[3]
0	3.1416	3.1416	6.2832	6.2832	9.4248	9.4248	1.566	1.566
0.1	3.0685	3.0685	5.7817	5.7817	8.0400	8.0400	9.9161	9.9161
0.3	2.6800	2.6800	4.3013	4.3013	5.4422	5.4422	6.3630	6.3630

Table 3.2
Validation of the Present Result with [3] for C-C Case

$\tau = \dfrac{e_0 a}{L}$	First Mode		Second Mode		Third Mode		Fourth Mode	
	Present	[3]	Present	[3]	Present	[3]	Present	[3]
0	4.7300	4.7300	7.8532	7.8532	10.9956	10.9956	14.1372	14.1372
0.1	4.5945	4.5945	7.1402	7.1402	9.2583	9.2583	11.016	11.016
0.3	3.9184	3.9184	5.1963	5.1963	6.2317	6.2317	7.0482	7.0482

Table 3.3
Validation of the Present Result with [4] for P-P Case

$(e_0 a)^2$	First Mode		Second Mode		Third Mode		Fourth Mode	
	Present	[4]	Present	[4]	Present	[4]	Present	[4]
0	3.1416	3.1416	6.2832	6.2832	9.4248	9.4248	12.5662	12.5664
1	3.0685	3.0685	5.7817	5.7817	8.0399	8.0400	9.9159	9.9161
2	3.0032	3.0032	5.4324	5.4324	7.3012	7.3012	8.7999	8.8000
3	2.9444	2.9444	5.1683	5.1683	6.8117	6.8118	8.1194	8.1195

Table 3.4
Validation of the Present Result with [4] for C-C Case

$(e_0 a)^2$	First Mode		Second Mode		Third Mode		Fourth Mode	
	Present	[4]	Present	[4]	Present	[4]	Present	[4]
0	4.7300	4.7300	7.8532	7.8532	10.9956	10.9956	14.1372	14.1358
1	4.5945	4.5945	7.1402	7.1403	9.2583	9.2583	11.0158	11.0138
2	4.4758	4.4758	6.6629	6.6629	8.3739	8.3739	9.7544	9.7519
3	4.3707	4.3707	6.3108	6.3108	7.8004	7.8004	8.9945	8.9916

agreement of present results with other well-known results found in previously published literature in special cases.

3.4.2 CONVERGENCE

To demonstrate the accuracy, efficiency, and powerfulness of the present analysis, convergence study is carried out for pinned–pinned (P-P) and clamped–clamped (C-C) edges. For the first case, i.e., "Young's modulus is imprecisely defined", we have taken $\alpha = 0.5$, $\beta = 0.5$, $\tau = 0.1$, and $L = 10$. Similarly, for the second case where "mass density is imprecisely defined", the parameters are taken as $\alpha = 0.5$, $\beta = 0.5$, $\tau = 0.5$, and $L = 10$. For the last case, i.e., when both the parameters E and ρ are uncertain, the scaling parameters are considered as $\alpha = 0.5$, $\beta = 0.5$, $\tau = 0.3$, and $L = 10$. The responses of the number of polynomials (n) on frequency parameters $\left(\sqrt{\lambda}\right)$ are noted through graphical and tabular results for the first three modes illustrated in Tables 3.5–3.7. From these results, it is witnessed that first-mode

Table 3.5
Convergence of Present Result When "Young's Modulus Is Imprecisely Defined"

(n)	Pinned–Pinned (P-P)			Clamped–Clamped (C-C)		
	First Mode	Second Mode	Third Mode	First Mode	Second Mode	Third Mode
3	3.0690	6.4905	9.6288	4.5945	7.2416	9.5951
4	3.0690	5.7934	9.6288	4.5945	7.1418	9.5951
5	3.0685	5.7934	8.1015	4.5945	7.1418	9.2703
6	3.0685	5.7817	8.1015	4.5945	7.1403	9.2703
7	3.0685	5.7817	8.0409	4.5945	7.1403	9.2585
8	3.0685	5.7817	8.0409	4.5945	7.1402	9.2585
9	3.0685	5.7817	8.0400	4.5945	7.1402	9.2583
10	3.0685	5.7817	8.0400	4.5945	7.1402	9.2583
11	3.0685	5.7817	8.0400	4.5945	7.1402	9.2583
12	3.0685	5.7817	8.0400	4.5945	7.1402	9.2583
13	3.0685	5.7817	8.0400	4.5945	7.1402	9.2583
14	3.0685	5.7817	8.0400	4.5945	7.1402	9.2583

Table 3.6
Convergence of Present Result When "Mass Density Is Imprecisely Defined"

(n)	Pinned–Pinned (P-P)			Clamped–Clamped (C-C)		
	First Mode	Second Mode	Third Mode	First Mode	Second Mode	Third Mode
3	2.3026	3.8475	5.0587	3.3156	4.2621	5.2076
4	2.3026	3.4668	5.0587	3.3156	4.1583	5.2076
5	2.3022	3.4668	4.3231	3.3153	4.1583	4.9463
6	2.3022	3.4604	4.3231	3.3153	4.1561	4.9463
7	2.3022	3.4604	4.2945	3.3153	4.1561	4.9330
8	2.3022	3.4604	4.2945	3.3153	4.1561	4.9330
9	2.3022	3.4604	4.2945	3.3153	4.1561	4.9328
10	2.3022	3.4604	4.2941	3.3153	4.1561	4.9328
11	2.3022	3.4604	4.2941	3.3153	4.1561	4.9328
12	2.3022	3.4604	4.2941	3.3153	4.1561	4.9328
13	2.3022	3.4604	4.2941	3.3153	4.1561	4.9328
14	2.3022	3.4604	4.2941	3.3153	4.1561	4.9328

Table 3.7

Convergence of Present Result When "Young's Modulus and Mass Density Are Imprecisely Defined"

(n)	Pinned–Pinned (P-P)			Clamped–Clamped (C-C)		
	First Mode	Second Mode	Third Mode	First Mode	Second Mode	Third Mode
3	2.6804	4.7917	6.4257	3.9185	5.3156	6.5454
4	2.6804	4.3094	6.4257	3.9185	5.1986	6.5454
5	2.6800	4.3094	5.4797	3.9184	5.1986	6.2458
6	2.6800	4.3014	5.4797	3.9184	5.1963	6.2458
7	2.6800	4.3014	5.4428	3.9184	5.1963	6.2319
8	2.6800	4.3013	5.4428	3.9184	5.1963	6.2319
9	2.6800	4.3013	5.4423	3.9184	5.1963	6.2317
10	2.6800	4.3013	5.4428	3.9184	5.1963	6.2317
11	2.6800	4.3013	5.4422	3.9184	5.1963	6.2317
12	2.6800	4.3013	5.4422	3.9184	5.1963	6.2317
13	2.6800	4.3013	5.4422	3.9184	5.1963	6.2317
14	2.6800	4.3013	5.4422	3.9184	5.1963	6.2317

frequency is converging faster with six terms as compared to other higher modes. But when the number of polynomials (n) approaches 9, frequency parameters for all the modes attain the convergence.

3.4.3 PROPAGATION OF UNCERTAINTY

In order to study the propagation of uncertainty in frequency parameters, three cases are examined. In the first case, Young's modulus is taken as fuzzy or imprecise whereas, in second and third cases, mass density and both the parameters (Young's modulus and mass density) are taken as input imprecise or fuzzy parameters, respectively. Tables 3.8(a–b) and Figures 3.1–3.2 depict the uncertainty in frequency parameters for both the lower and upper bounds of P-P and C-C boundary conditions for the first case. Likewise, Tables 3.9(a–b) and Figures 3.3–3.4 demonstrate the propagation of uncertainty in frequency parameters in terms of lower and upper bounds for the second case whereas Tables 3.10(a–b) and Figures 3.5–3.6 represent the uncertainty propagation of the third case. From the graphical results, it is interesting to note that by taking $\alpha = 0$, we get the frequency parameters in interval forms with both the lower and upper bounds. Also, by considering $\alpha = 1$, the frequency parameters for the deterministic (or exact or crisp) case are obtained. It may also be noted that $\beta = 0$ gives the lower bound, whereas $\beta = 1$ gives the upper bound of the frequency parameters. It is also interesting to observe that by taking $\alpha = 1$, both the upper and lower bounds give the same results, which are the same as deterministic (or crisp) value. In all the above cases, the computations are carried out by considering the nonlocal parameter (τ) as 0.5.

Table 3.8
Lower and Upper Bound Frequencies When "Young's Modulus Is Imprecisely Defined"

(a) For Pinned-Pinned (P-P) Case

	$\beta=0$				$\beta=1$			
α	$\sqrt{\lambda_1}$	$\sqrt{\lambda_2}$	$\sqrt{\lambda_3}$	$\sqrt{\lambda_4}$	$\sqrt{\lambda_1}$	$\sqrt{\lambda_2}$	$\sqrt{\lambda_3}$	$\sqrt{\lambda_4}$
0	1.9359	2.9098	3.6109	4.1894	2.5478	3.8296	4.7522	5.5135
0.1	1.9826	2.9800	3.6979	4.2904	2.5263	3.7972	4.7121	5.4670
0.2	2.0262	3.0455	3.7793	4.3847	2.5043	3.7641	4.6709	5.4192
0.3	2.0672	3.1071	3.8556	4.4734	2.4816	3.7300	4.6286	5.3702
0.4	2.1058	3.1652	3.9277	4.5570	2.4583	3.6950	4.5852	5.3198
0.5	2.1425	3.2203	3.9961	4.6363	2.4343	3.6589	4.5404	5.2679
0.6	2.1773	3.2726	4.0611	4.7117	2.4096	3.6218	4.4943	5.2144
0.7	2.2106	3.3226	4.1231	4.7837	2.3841	3.5834	4.4468	5.1592
0.8	2.2424	3.3704	4.1824	4.8525	2.3577	3.5438	4.3976	5.1022
0.9	2.2729	3.4163	4.2394	4.9186	2.3305	3.5029	4.3468	5.0432
1.0	2.3022	3.4604	4.2941	4.9820	2.3022	3.4604	4.2941	4.9820

(b) For Clamped–Clamped (C-C) Case

	$\beta=0$				$\beta=1$			
α	$\sqrt{\lambda_1}$	$\sqrt{\lambda_2}$	$\sqrt{\lambda_3}$	$\sqrt{\lambda_4}$	$\sqrt{\lambda_1}$	$\sqrt{\lambda_2}$	$\sqrt{\lambda_3}$	$\sqrt{\lambda_4}$
0	2.7878	3.4948	4.1480	4.6428	3.6690	4.5995	5.4590	6.1103
0.1	2.8551	3.5791	4.2480	4.7548	3.6380	4.5606	5.4130	6.0588
0.2	2.9179	3.6578	4.3414	4.8594	3.6063	4.5208	5.3657	6.0058
0.3	2.9768	3.7317	4.4292	4.9576	3.5736	4.4799	5.3171	5.9515
0.4	3.0325	3.8015	4.5120	5.0503	3.5401	4.4378	5.2672	5.8956
0.5	3.0853	3.8677	4.5905	5.1382	3.5055	4.3945	5.2158	5.8381
0.6	3.1354	3.9306	4.6651	5.2217	3.4699	4.3499	5.1628	5.7788
0.7	3.1833	3.9906	4.7364	5.3015	3.4332	4.3039	5.1082	5.7176
0.8	3.2291	4.0480	4.8046	5.3778	3.3953	4.2563	5.0517	5.6544
0.9	3.2731	4.1031	4.8699	5.4510	3.3560	4.2071	4.9933	5.5891
1.0	3.3153	4.1561	4.9328	5.5213	3.3153	4.1561	4.9328	5.5213

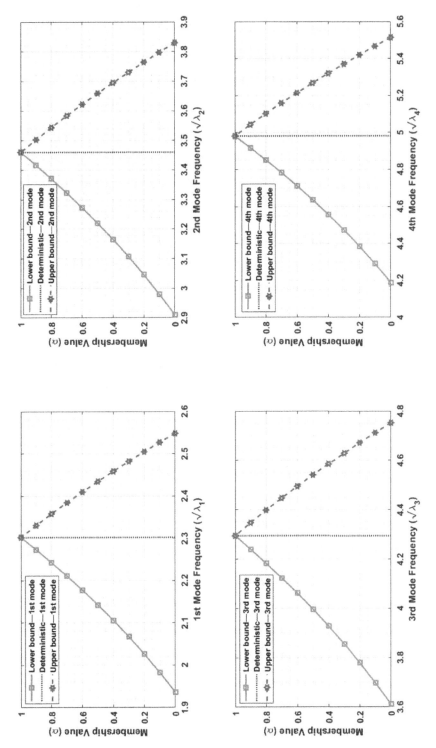

FIGURE 3.1 Triangular Fuzzy Number (TFN) when E is imprecise for P-P case.

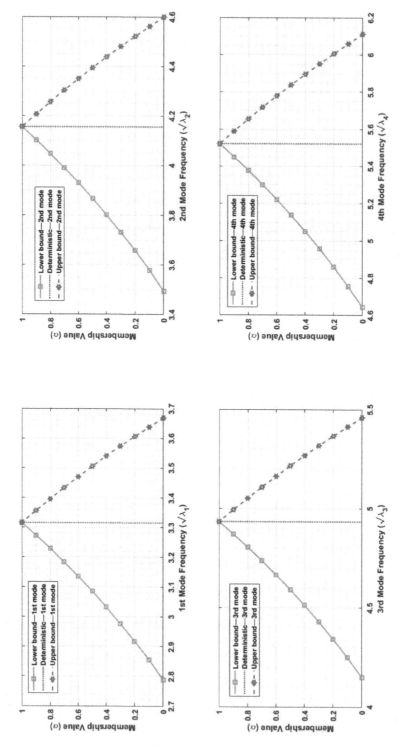

FIGURE 3.2 Triangular Fuzzy Number (TFN) when E is imprecise for C–C case.

Table 3.9
Lower and Upper Bound Frequencies When "Mass Density Is Imprecisely Defined"
(a) For Pinned-Pinned (P-P) Case

α	$\beta=0$				$\beta=1$			
	$\sqrt{\lambda_1}$	$\sqrt{\lambda_2}$	$\sqrt{\lambda_3}$	$\sqrt{\lambda_4}$	$\sqrt{\lambda_1}$	$\sqrt{\lambda_2}$	$\sqrt{\lambda_3}$	$\sqrt{\lambda_4}$
0	2.4343	3.6589	4.5404	5.2679	2.1996	3.3062	4.1027	4.7601
0.1	2.4193	3.6364	4.5125	5.2354	2.2089	3.3201	4.1200	4.7801
0.2	2.4048	3.6146	4.4854	5.2040	2.2184	3.3344	4.1377	4.8006
0.3	2.3907	3.5934	4.4591	5.1735	2.2280	3.3489	4.1557	4.8215
0.4	2.3770	3.5728	4.4335	5.1438	2.2379	3.3637	4.1741	4.8429
0.5	2.3637	3.5528	4.4087	5.1150	2.2480	3.3789	4.1930	4.8647
0.6	2.3507	3.5333	4.3845	5.0870	2.2584	3.3945	4.2122	4.8871
0.7	2.3381	3.5143	4.3610	5.0597	2.2689	3.4104	4.2320	4.9100
0.8	2.3258	3.4959	4.3381	5.0331	2.2798	3.4266	4.2522	4.9334
0.9	2.3139	3.4779	4.3158	5.0073	2.2909	3.4433	4.2729	4.9574
1.0	2.3022	3.4604	4.2941	4.9820	2.3022	3.4604	4.2941	4.9820

(b) For Clamped–Clamped (C-C) Case

α	$\beta=0$				$\beta=1$			
	$\sqrt{\lambda_1}$	$\sqrt{\lambda_2}$	$\sqrt{\lambda_3}$	$\sqrt{\lambda_4}$	$\sqrt{\lambda_1}$	$\sqrt{\lambda_2}$	$\sqrt{\lambda_3}$	$\sqrt{\lambda_4}$
0	3.5055	4.3945	5.2158	5.8381	3.1676	3.9709	4.7130	5.2753
0.1	3.4840	4.3675	5.1837	5.8021	3.1809	3.9876	4.7328	5.2975
0.2	3.4630	4.3412	5.1526	5.7673	3.1946	4.0047	4.7531	5.3202
0.3	3.4427	4.3158	5.1223	5.7335	3.2085	4.0221	4.7738	5.3434
0.4	3.4230	4.2910	5.0930	5.7006	3.2227	4.0400	4.7950	5.3671
0.5	3.4038	4.2670	5.0644	5.6687	3.2373	4.0582	4.8166	5.3913
0.6	3.3852	4.2436	5.0367	5.6376	3.2521	4.0769	4.8388	5.4161
0.7	3.3670	4.2209	5.0097	5.6074	3.2674	4.0960	4.8615	5.4415
0.8	3.3493	4.1987	4.9834	5.5779	3.2830	4.1155	4.8847	5.4674
0.9	3.3321	4.1771	4.9578	5.5493	3.2990	4.1356	4.9084	5.4940
1.0	3.3153	4.1561	4.9328	5.5213	3.3153	4.1561	4.9328	5.5213

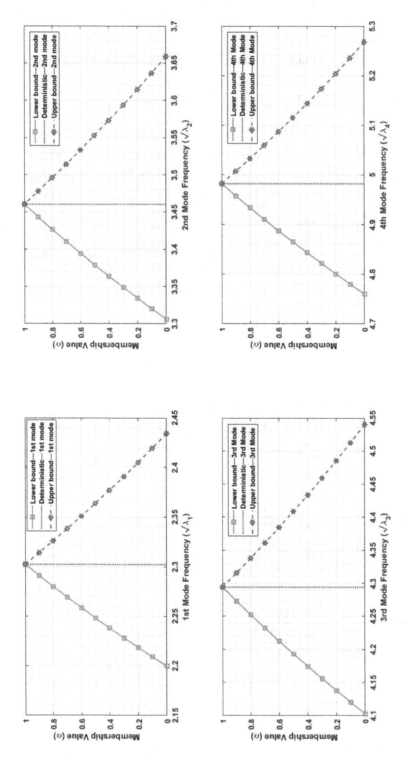

FIGURE 3.3 Triangular Fuzzy Number (TFN) when ρ is imprecise for P-P case.

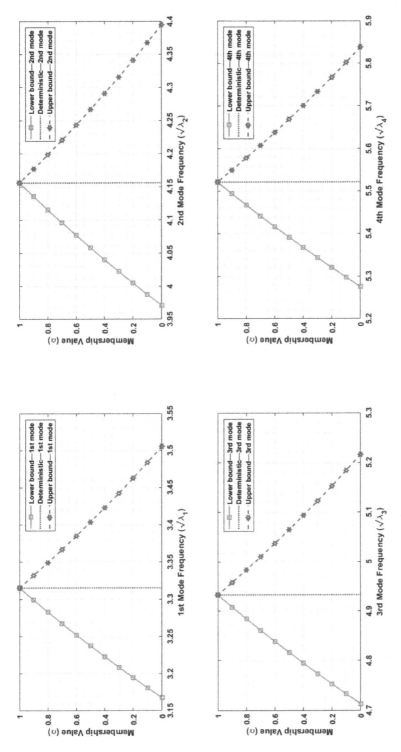

FIGURE 3.4 Triangular Fuzzy Number (TFN) when ρ is imprecise for C-C case.

Table 3.10
Lower and Upper Bound Frequencies When "Young's Modulus and Mass Density Are Imprecisely Defined"

(a) For Pinned–Pinned (P-P) Case

	$\beta = 0$				$\beta = 1$			
α	$\sqrt{\lambda_1}$	$\sqrt{\lambda_2}$	$\sqrt{\lambda_3}$	$\sqrt{\lambda_4}$	$\overline{\sqrt{\lambda_1}}$	$\overline{\sqrt{\lambda_2}}$	$\overline{\sqrt{\lambda_3}}$	$\overline{\sqrt{\lambda_4}}$
0	2.0470	3.0768	3.8180	4.4297	2.4343	3.6589	4.5404	5.2679
0.1	2.0835	3.1316	3.8860	4.5086	2.4239	3.6433	4.5211	5.2454
0.2	2.1165	3.1812	3.9476	4.5801	2.4131	3.6270	4.5008	5.2219
0.3	2.1466	3.2265	4.0038	4.6453	2.4016	3.6098	4.4795	5.1971
0.4	2.1742	3.2680	4.0553	4.7050	2.3896	3.5918	4.4571	5.1712
0.5	2.1996	3.3062	4.1027	4.7601	2.3770	3.5728	4.4335	5.1438
0.6	2.2232	3.3416	4.1466	4.8110	2.3637	3.5528	4.4087	5.1150
0.7	2.2450	3.3744	4.1874	4.8582	2.3496	3.5316	4.3824	5.0846
0.8	2.2654	3.4050	4.2253	4.9023	2.3347	3.5093	4.3547	5.0524
0.9	2.2844	3.4336	4.2608	4.9435	2.3190	3.4856	4.3253	5.0183
1.0	2.3022	3.4604	4.2941	4.9820	2.3022	3.4604	4.2941	4.9820

(b) For Clamped–Clamped (C-C) Case

	$\beta = 0$				$\beta = 1$			
α	$\sqrt{\lambda_1}$	$\sqrt{\lambda_2}$	$\sqrt{\lambda_3}$	$\sqrt{\lambda_4}$	$\overline{\sqrt{\lambda_1}}$	$\overline{\sqrt{\lambda_2}}$	$\overline{\sqrt{\lambda_3}}$	$\overline{\sqrt{\lambda_4}}$
0	2.9478	3.6953	4.3859	4.9092	3.5055	4.3945	5.2158	5.8381
0.1	3.0003	3.7612	4.4641	4.9966	3.4906	4.3758	5.1935	5.8132
0.2	3.0479	3.8208	4.5348	5.0759	3.4749	4.3561	5.1702	5.7871
0.3	3.0912	3.8751	4.5993	5.1481	3.4585	4.3355	5.1458	5.7597
0.4	3.1310	3.9250	4.6585	5.2143	3.4412	4.3138	5.1200	5.7309
0.5	3.1676	3.9709	4.7130	5.2753	3.4230	4.2910	5.0930	5.7006
0.6	3.2015	4.0134	4.7634	5.3317	3.4038	4.2670	5.0644	5.6687
0.7	3.2329	4.0528	4.8102	5.3841	3.3836	4.2416	5.0343	5.6349
0.8	3.2623	4.0896	4.8538	5.4329	3.3621	4.2148	5.0024	5.5993
0.9	3.2897	4.1239	4.8946	5.4786	3.3394	4.1863	4.9687	5.5615
1.0	3.3153	4.1561	4.9328	5.5213	3.3153	4.1561	4.9328	5.5213

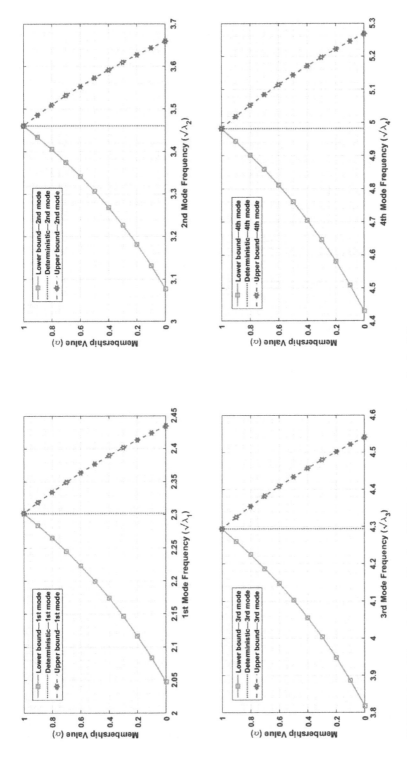

FIGURE 3.5 Triangular Fuzzy Number (TFN) when E and ρ are imprecise for P-P case.

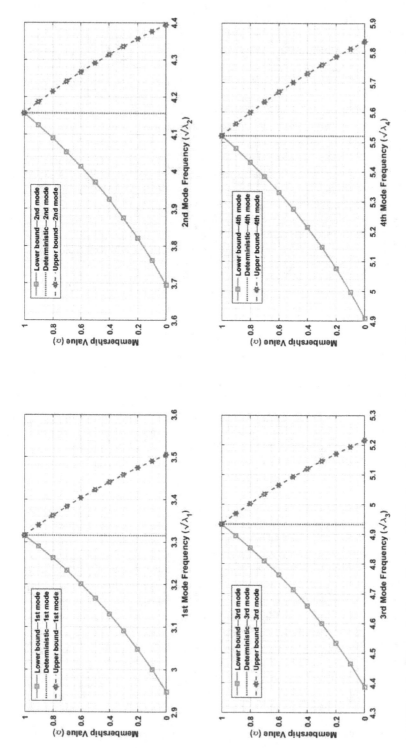

FIGURE 3.6 Triangular Fuzzy Number (TFN) when E and ρ are imprecise for C–C case.

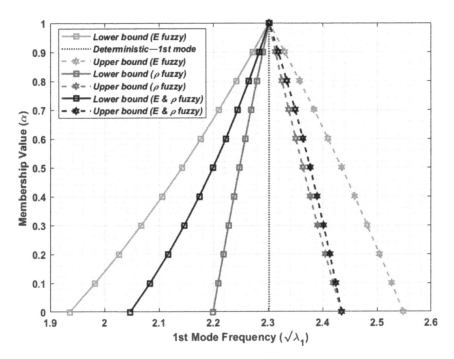

FIGURE 3.7   Triangular Fuzzy Number (TFN) for $\sqrt{\lambda_1}$ P-P case.

Again, all the uncertain frequencies for the above three cases are compared, which are illustrated in Figures 3.7–3.8. Figure 3.7 represents a fundamental uncertain frequency parameter for P-P edge, and Figure 3.8 is for fundamental uncertain frequency parameter for C-C edge. From both the figures, it is very evident that comparatively larger width is achieved for both the cases, i.e., P-P and C-C edges when the uncertainty or fuzziness is taken along Young's modulus. Similarly, the lower width has appeared in the figures when mass density is fuzzy or imprecise.

3.4.4   MODE SHAPES

Mode shapes are very crucial for studying the dynamic behavior of structural elements. Also, mode shapes are susceptible to uncertainty. As regards, mode shapes are plotted for the above three cases which are depicted in Figures 3.9–3.11. In Figure 3.9, a comparative study is carried out between all the first four modes of P-P and C-C boundary conditions. This graphical result is plotted with $\tau = 0.5$, $\alpha = \beta = 0.5$, and uncertainty is taken in Young's modulus. Similarly, Figure 3.10 and Figure 3.11 are illustrated for second and third cases, i.e., Figure 3.10 is plotted by considering mass density as fuzzy with $\tau = 0.3$ and $\alpha = \beta = 0.5$, whereas Figure 3.11 represents the mode shapes which are plotted with $\tau = 0.1$ and $\alpha = \beta = 0.5$ for both the uncertain parameters i.e., Young's modulus as well as mass density.

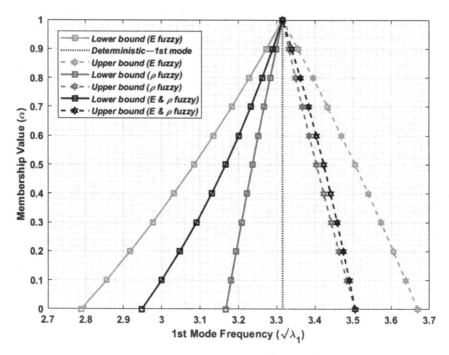

FIGURE 3.8   Triangular Fuzzy Number (TFN) for $\sqrt{\lambda_1}$ for C-C case.

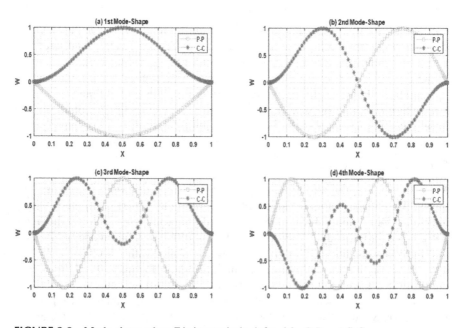

FIGURE 3.9   Mode-shape when E is imprecisely defined for P-P and C-C cases.

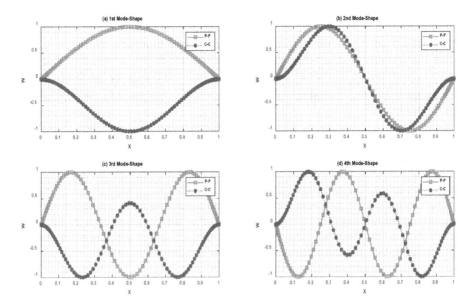

FIGURE 3.10   Mode-shape when ρ is imprecisely defined for P-P and C-C cases.

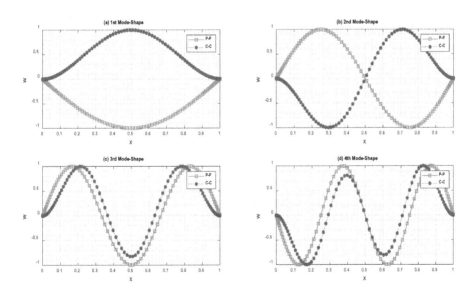

FIGURE 3.11   Mode-shape when E and ρ are imprecisely defined for P-P and C-C cases.

3.5 CONCLUDING REMARKS

This chapter delves into the modeling of imprecisely defined Euler–Bernoulli nanobeams within the framework of Eringen's nonlocal elasticity theory. The study extensively investigates the propagation of uncertainty in frequency parameters for both pinned–pinned (P-P) and clamped–clamped (C-C) boundary conditions. Key highlights include:

- The investigation encompasses three distinct cases: (i) Imprecisely defined Young's modulus, (ii) imprecisely defined mass density, and (iii) simultaneous imprecision in both Young's modulus and mass density. Uncertainty is characterized using Triangular Fuzzy Numbers.
- The transverse displacement equation for the imprecisely defined Euler–Bernoulli nanobeam is reformulated into a double parametric form to address uncertainties. Subsequently, the Rayleigh–Ritz method is employed to compute uncertain frequency parameters with respect to varying values of the parameters alpha and beta.
- Graphical results illustrate that uncertainty in Young's modulus exhibits a broader range in frequency parameters for both P-P and C-C boundary conditions, whereas the range is comparatively narrower when mass density is imprecise.
- The accuracy and efficiency of the models are validated through convergence studies for both P-P and C-C boundaries. The influence of the number of polynomials on frequency parameters for the first three modes is explored, revealing faster convergence for the first mode compared to higher modes, with convergence achieved for all modes when the number of polynomials reaches nine.
- In uncertain scenarios, lower and upper bounds of frequency parameters are computed, and graphical representations are generated as Triangular Fuzzy Numbers to demonstrate the models' sensitivity to material uncertainties. Additionally, mode shapes are plotted for the uncertain cases corresponding to the first four frequency parameters.

BIBLIOGRAPHY

[1] Reddys J N 2007 Nonlocal theories for bending, buckling and vibration of beams *International Journal Engineering Science* 45(2) 288–307

[2] Eringen A C 1972 Nonlocal polar elastic continua *International Journal of Engineering Science* 10 1–16

[3] Wang C, Zhang Y and He X 2007 Vibration of nonlocal Timoshenko beams *Nanotechnology* 18 105–113

[4] Jena S K and Chakraverty S 2018 Free vibration analysis of Euler–Bernoulli nanobeam using differential transform method *International Journal of Computational Materials Science and Engineering* 7 1850020

4 Stability Analysis of Euler–Bernoulli Nanobeam with Uncertainty

4.1 INTRODUCTION

In this chapter, a non-probabilistic approach-based Navier's method (NM) and Galerkin Weighted Residual Method (GWRM) in term of double parametric form have been proposed to investigate the buckling behavior of Euler–Bernoulli nonlocal beam under the framework of the Eringen's nonlocal elasticity theory, considering the structural parameters as imprecise or uncertain. The uncertainties in Young's modulus and diameter of the beam are modeled in terms of Triangular Fuzzy Numbers (TFNs). The critical buckling loads are calculated for hinged–hinged (HH), clamped–hinged (CH), and clamped–clamped (CC) boundary conditions and these results are compared with the deterministic model in special cases, demonstrating robust agreement. Further, a random sampling technique-based method, namely, Monte Carlo Simulation Technique (MCST) has been implemented to compute the critical buckling loads of uncertain systems. Also, the critical buckling loads obtained from the uncertain model in terms of lower bound (LB) and upper bound (UB) by the non-probabilistic methods, namely Navier's method (NM) and Galerkin Weighted Residual Method (GWRM), are again verified with the Monte Carlo Simulation Technique (MCST) with their time periods, demonstrating the efficacy, accuracy, and effectiveness of the proposed uncertain model. A comparative study is also carried out among the non-probabilistic methods and Monte Carlo Simulation Technique (MCST) to demonstrate the effectiveness of methods with respect to time. Additionally, a parametric study has been performed to display the propagation of uncertainties into the nonlocal system in the form of critical buckling loads.

4.2 FORMULATION OF THE PROPOSED MODEL

In this section, the governing equation for the stability analysis of the nanobeam has been developed by considering the material uncertainties associated with Young's modulus and the diameter of the nanobeam in terms of Triangular Fuzzy Number. The double parametric form is also implemented to derive the governing equation with structural uncertainties.

DOI: 10.1201/9781003303107-4

4.2.1 Governing Equations of Motion

The equation of motion for the buckling of Euler–Bernoulli beam in term of bending moment and transverse displacement may be expressed as Eq. (4.1):

$$\frac{d^2 M}{dx^2} = \hat{P} \frac{d^2 w}{dx^2} \tag{4.1}$$

In which, M is the bending moment and $\hat{P}$ is the applied compressive force due to mechanical load. The nonlocal constitutive relation by implementing Eringen's nonlocal elasticity theory [2] may be expressed as [1]:

$$M^{nl} - (e_0 a)^2 \frac{d^2 M}{dx^2} = -EI \frac{d^2 w}{dx^2} \tag{4.2}$$

Where E, I, and $(e_0 a)^2$ represent Young's modulus, second moment of area, and small-scale parameter of the beam, respectively.

Using Eq. (4.1) into Eq. (4.2) and simplifying, we obtain:

$$M^{nl} = -EI \frac{d^2 w}{dx^2} + (e_0 a)^2 \left(\hat{P} \frac{d^2 w}{dx^2} \right) \tag{4.3}$$

4.2.2 Modeling with Material Uncertainties

In this chapter, both Young's modulus and the diameter of the beam are considered as uncertain or random in terms of Triangular Fuzzy Number (TFN). Therefore, the uncertain Young's modulus and diameter can be expressed as:

$$\tilde{E} = (e_1, e_2, e_3) \qquad \text{and} \qquad \tilde{d} = (d_1, d_2, d_3) \tag{4.4}$$

In which d_2 and e_2 represent deterministic values of Young's modulus and diameter of the beam, respectively. The graphical representations of the uncertain Young's modulus and diameter are demonstrated in Figure 4.1 and Figure 4.2, respectively.

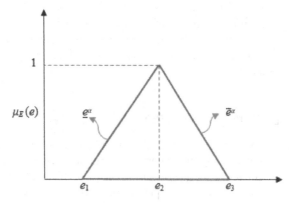

FIGURE 4.1 Graphical representation of $\tilde{E} = (e_1, e_2, e_3)$.

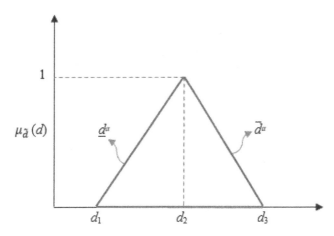

FIGURE 4.2 Graphical representation of $\tilde{d} = (d_1, d_2, d_3)$.

Plugging Eq. (4.4) into Eq. (4.3), we have:

$$M^{nl} = -\tilde{E}\frac{\pi \tilde{d}^4}{64}\frac{d^2 w}{dx^2} + (e_0 a)^2 \left[\hat{P}\frac{d^2 w}{dx^2}\right] \tag{4.5}$$

Using Eq. (4.5) in Eq. (4.1), the governing equation of motion for stability analysis of nonlocal beam considering the uncertainties in Young's modulus and diameter can be expressed as:

$$-\tilde{E}\frac{\pi \tilde{d}^4}{64}\frac{d^4 w}{dx^4} + (e_0 a)^2 \hat{P}\frac{d^4 w}{dx^4} - \hat{P}\frac{d^2 w}{dx^2} = 0 \tag{4.6}$$

Incorporating the concept of the double parametric form as given in preliminaries, the Triangular Fuzzy Numbers $\tilde{E} = (e_1, e_2, e_3)$ and $\tilde{d} = (d_1, d_2, d_3)$ can be expressed in double parametric form as:

$$\tilde{E}(\alpha_1, \beta_1) = (e_1 - e_3)\alpha_1\beta_1 + (e_3 - e_1)\beta_1 + (e_2 - e_1)\alpha_1 + e_1, \alpha_1, \beta_1 \in [0\ 1] \tag{4.7}$$

$$\tilde{d}(\alpha_2, \beta_2) = (d_1 - d_3)\alpha_2\beta_2 + (d_3 - d_1)\beta_2 + (d_2 - d_1)\alpha_2 + d_1, \alpha_2, \beta_2 \in [0\ 1] \tag{4.8}$$

Plugging Eq. (4.7) and Eq. (4.8) in Eq. (4.6), the governing Eq. (4.6) can be modified into:

$$-\tilde{E}(\alpha_1, \beta_1)\frac{\pi}{64}\left(\tilde{d}(\alpha_2, \beta_2)\right)^4\frac{d^4 w}{dx^4} + (e_0 a)^2 \hat{P}\frac{d^4 w}{dx^4} - \hat{P}\frac{d^2 w}{dx^2} = 0 \tag{4.9}$$

4.3 SOLUTION PROCEDURES

In this chapter, three methods such as Navier's method, Galerkin weighted residual method, and Monte Carlo Simulation Technique have been utilized to find out the critical buckling loads of Euler–Bernoulli nanobeam having structural uncertainties.

Navier's method has been used for hinged–hinged (HH) boundary condition whereas Galerkin weighted residual method has been employed for clamped–hinged (CH) and clamped–clamped (CC) boundary conditions, based on the double parametric approach of Triangular Fuzzy Number. Monte Carlo Simulation Technique is also used for all the boundary conditions after converting the Triangular Fuzzy Number into interval form.

4.3.1 NAVIER'S METHOD

Navier's approach has been employed to compute the critical buckling loads of nano-beam with structural uncertainties analytically for hinged–hinged (HH) boundary condition. The transverse displacement (w) may be expressed as [3–5]:

$$w(x) = \sum_{n=1}^{\infty} W_n \sin\left(\frac{n\pi}{L}x\right) \tag{4.10}$$

In which W_n is the unknown to be determined. Plugging Eq. (4.10) into Eq. (4.9), the buckling load $\left(\hat{P}_n\right)$ may be obtained as:

$$\hat{P}_n = \frac{\tilde{E}(\alpha_1,\beta_1)\dfrac{\pi}{64}\left(\tilde{d}(\alpha_2,\beta_2)\right)^4\left(\dfrac{n\pi}{L}\right)^4}{\left(\dfrac{n\pi}{L}\right)^2 + (e_0 a)^2\left(\dfrac{n\pi}{L}\right)^4} \tag{4.11}$$

Now, by using the double parametric form for Young's modulus and diameter, the above equation will be modified into:

$$\hat{P}_n = \frac{\begin{bmatrix}\left((e_1-e_3)\alpha_1\beta_1+(e_3-e_1)\beta_1+(e_2-e_1)\alpha_1+e_1\right)\times \\ \left((d_1-d_3)\alpha_2\beta_2+(d_3-d_1)\beta_2+(d_2-d_1)\alpha_2+d_1\right)^4\end{bmatrix}\left(\dfrac{\pi}{64}\right)\left(\dfrac{n\pi}{L}\right)^4}{\left(\dfrac{n\pi}{L}\right)^2 + (e_0 a)^2\left(\dfrac{n\pi}{L}\right)^4} \tag{4.12}$$

Putting $n = 1, 2, 3$, etc., critical buckling load, second buckling load, third buckling load, etc. will be obtained.

4.3.2 GALERKIN WEIGHTED RESIDUAL METHOD

In this section, a double parametric form-based Galerkin weighted residual method has been used to compute critical buckling loads for clamped–hinged (CH) and clamped–clamped (CC) boundary conditions. For the exact solution, the right-hand side of Eq. (4.9) is zero, whereas the right-hand side of the Eq. (4.9) is not exactly zero in case of the approximate solution. So, the residue of the Eq. (4.9) can be defined as:

$$\Re = -\tilde{E}(\alpha_1,\beta_1)\frac{\pi}{64}\left(\tilde{d}(\alpha_2,\beta_2)\right)^4\frac{d^4 w}{dx^4} + (e_0 a)^2\,\hat{P}\frac{d^4 w}{dx^4} - \hat{P}\frac{d^2 w}{dx^2} \tag{4.13}$$

The transverse displacement function w is approximated by a weighting function Φ_n as:

$$w = \sum_{n=1}^{\infty} w_n \Phi_n(x) \tag{4.14}$$

Here, w_n is the unknown coefficient, and the function $\Phi_n(x)$ has different values for different boundary conditions which are given in the next few sections [6–7].

4.3.2.1 Clamped–Hinged (CH)

$$\Phi_n(x) = \left\langle \begin{array}{c} \sin\left(\dfrac{(n+0.25)\pi}{L}x\right) - \sinh\left(\dfrac{(n+0.25)\pi}{L}x\right) - \\[2ex] \xi_n\left[\cos\left(\dfrac{(n+0.25)\pi}{L}x\right) - \cosh\left(\dfrac{(n+0.25)\pi}{L}x\right)\right] \end{array} \right\rangle \tag{4.15}$$

where $\xi_n = \left|\dfrac{\sin(n+0.25)\pi + \sinh(n+0.25)\pi}{\cos(n+0.25)\pi + \cosh(n+0.25)\pi}\right|$

4.3.2.2 Clamped–Clamped (CC)

$$\Phi_n(x) = \left\langle \begin{array}{c} \sin\left(\dfrac{(n+0.5)\pi}{L}x\right) - \sinh\left(\dfrac{(n+0.5)\pi}{L}x\right) - \\[2ex] \xi_n\left[\cos\left(\dfrac{(n+0.5)\pi}{L}x\right) - \cosh\left(\dfrac{(n+0.5)\pi}{L}x\right)\right] \end{array} \right\rangle \tag{4.16}$$

where $\xi_n = \left|\dfrac{\sin(n+0.5)\pi - \sinh(n+0.5)\pi}{\cos(n+0.5)\pi - \cosh(n+0.5)\pi}\right|$

Multiplying weighting function $\Phi_n(x)$ with Eq. (4.13) and integrating the weighted residual over the length and letting the integral to zero lead to:

$$\int_0^L \Re\Phi_n(x)dx = 0 \Rightarrow \int_0^L \left[\begin{array}{c} -\tilde{E}(\alpha_1,\beta_1)\dfrac{\pi}{64}\left(\tilde{d}(\alpha_2,\beta_2)\right)^4 \dfrac{d^4w}{dx^4} \\[2ex] +\left(e_0a\right)^2 \hat{P}\dfrac{d^4w}{dx^4} - \hat{P}\dfrac{d^2w}{dx^2} \end{array} \right] \Phi_n(x)dx = 0 \tag{4.17}$$

Now by performing integration by parts, Eq. (4.17) will be modified into the weak-form governing differential equation as:

$$\int_0^L \left[\begin{array}{l} -\tilde{E}(\alpha_1,\beta_1)\dfrac{\pi}{64}\left(\tilde{d}(\alpha_2,\beta_2)\right)^4 \dfrac{d^2\Phi_n(x)}{dx^2}\dfrac{d^2\Phi_n(x)}{dx^2} \\ +\left(e_0a\right)^2 \hat{P}\dfrac{d^2\Phi_n(x)}{dx^2}\dfrac{d^2\Phi_n(x)}{dx^2}+\hat{P}\dfrac{d\Phi_n(x)}{dx}\dfrac{d\Phi_n(x)}{dx} \end{array} \right] dx = 0 \qquad (4.18)$$

Simplifying the above Eq. (4.18), buckling loads $\left(\hat{P}\right)$ of the nonlocal beam can be obtained as:

$$\hat{P}_n = \frac{\displaystyle\int_0^L \tilde{E}(\alpha_1,\beta_1)\dfrac{\pi}{64}\left(\tilde{d}(\alpha_2,\beta_2)\right)^4 \dfrac{d^2\Phi_n(x)}{dx^2}\dfrac{d^2\Phi_n(x)}{dx^2}}{\displaystyle\int_0^L \dfrac{d\Phi_n(x)}{dx}\dfrac{d\Phi_n(x)}{dx}+\left(e_0a\right)^2\dfrac{d^2\Phi_n(x)}{dx^2}\dfrac{d^2\Phi_n(x)}{dx^2}} \qquad (4.19)$$

Using the double parametric form as given in Eqs. (4.7–4.8), Eq. (4.19) will be changed into:

$$\hat{P}_n = \frac{\displaystyle\int_0^L \left[\begin{array}{l}\left((e_1-e_3)\alpha_1\beta_1+(e_3-e_1)\beta_1+(e_2-e_1)\alpha_1+e_1\right)\times \\ \left((d_1-d_3)\alpha_2\beta_2+(d_3-d_1)\beta_2+(d_2-d_1)\alpha_2+d_1\right)^4\end{array}\right]\left(\dfrac{\pi}{64}\right)\dfrac{d^2\Phi_n(x)}{dx^2}\dfrac{d^2\Phi_n(x)}{dx^2}}{\displaystyle\int_0^L \dfrac{d\Phi_n(x)}{dx}\dfrac{d\Phi_n(x)}{dx}+\left(e_0a\right)^2\dfrac{d^2\Phi_n(x)}{dx^2}\dfrac{d^2\Phi_n(x)}{dx^2}} \qquad (4.20)$$

4.3.3 Monte Carlo Simulation Technique

The uncertain parameters in terms of Triangular Fuzzy Numbers (TFNs) are converted into interval form by using a single parametric form as presented in preliminaries, namely α cut technique as:

$$\tilde{E}=(e_1,e_2,e_3)=\left[\underline{E}(\alpha_1)\ \overline{E}(\alpha_1)\right]=\left[(e_2-e_1)\alpha_1+e_1\ -(e_3-e_2)\alpha_1+e_3\right],$$
$$\alpha_1 \in \begin{bmatrix}0 & 1\end{bmatrix} \qquad (4.21)$$

$$\tilde{d}=(d_1,d_2,d_3)=\left[\underline{d}(\alpha_2)\ \overline{d}(\alpha_2)\right]=\left[(d_2-d_1)\alpha_2+d_1\ -(d_3-d_2)\alpha_2+d_3\right],$$
$$\alpha_2 \in \begin{bmatrix}0 & 1\end{bmatrix} \qquad (4.22)$$

Random points numbering 10,000 have been generated from these intervals, and these random points of uncertain parameters are used in the deterministic analysis of critical buckling load. By repeating the deterministic analysis for all the random points, lower bound and upper bound and mean of critical buckling loads are computed. The mean of the critical buckling loads, obtained by the Monte Carlo Simulation Technique, represents the results of the deterministic model.

4.4   NUMERICAL RESULTS AND DISCUSSION

In this study, critical buckling loads (P_{cr}) of nonlocal beam have been computed for HH, CH, and CC boundary conditions by developing MATLAB codes for the proposed uncertain model. The study has progressed by considering three cases: (i) Young's modulus (E) is uncertain or fuzzy, (ii) diameter (d) is uncertain or fuzzy, and (iii) Young's modulus (E) and diameter (d) are uncertain or fuzzy. For computational purposes, the structural parameters are considered as $\tilde{E} = (0.6, 1, 1.4)$ TPa, where $E = 1$ TPa is the deterministic value, $\tilde{d} = (0.8, 1, 1.2)$ nm, where $d = 1$ nm is the deterministic value, $L = 1$ nm, and $I = \dfrac{\pi \tilde{d}^4}{64}$.

4.4.1   VALIDATION OF THE PROPOSED MODEL

The proposed uncertain model is validated by comparing the critical buckling loads with [8] by considering the deterministic values for Young's modulus and diameter. It may be noted that by considering $\alpha_1 = \alpha_2 = 1$, the uncertain models turn into a deterministic model. Critical buckling loads are compared for HH boundary condition by using Navier's method (NM) and Monte Carlo Simulation Technique (MCST) which are depicted in Table 4.1 as tabular results. From this result, a robust agreement between the present results obtained by both the non-probabilistic method and Monte Carlo Simulation Technique and the existing results of [8] is witnessed.

4.4.2   PROPAGATION OF UNCERTAINTIES IN CRITICAL BUCKLING LOADS

Through this subsection, the propagation of structural uncertainties into dynamical characteristics, in particular, critical buckling loads, has been studied comprehensively. In order to proceed smoothly, three cases have been considered, which are the systems having (i) Young's modulus (E) as uncertain parameters, (ii) diameter (d) as uncertain parameters, and finally (iii) both Young's modulus (E) and diameter (d) as uncertain parameters. It may be noted that double parametric form-based Navier's method (NM) has been used for HH boundary condition whereas Galerkin Weighted Residual Method has been employed for CH and CC boundary conditions. Also, the random sampling technique-based Monte Carlo Simulation Technique is used to compute and verify the results.

Table 4.1

Comparisons of P_{cr} (in nN) for HH Boundary Condition

$(e_0 a)$ in nm	0	0.5	1	1.5	2
P_{cr} - NM (Present)	4.8447	4.7281	4.4095	3.9644	3.4735
P_{cr} - MCST (Present)	4.8447	4.7281	4.4095	3.9644	3.4735
P_{cr} [8]	4.8447	4.7281	4.4095	3.9644	3.4735

The tabular and graphical results for case (i) have been presented in Table 4.2 and Figures 4.3–4.5, respectively, considering the structural parameters as $\tilde{E} = (0.6, 1, 1.4)\,\text{TPa}$, $d = 1\,nm, e_0 a = 1\,nm$, and $L = 10\,nm$. The critical buckling loads of the uncertain system are being computed through lower bound and upper

---

Table 4.2

LB and UB of Critical Buckling Loads P_{cr} in nN, Considering Young's Modulus as an Uncertain Parameter with $d = 1$ nm, $e_0 a = 1$ nm, and $L = 10$ nm.

(a) Hinged–Hinged (HH) Boundary Condition

α_1	LB of Critical Buckling Load $\left(\underline{P}_{cr}\right)$			UB of Critical Buckling Load $\left(\bar{P}_{cr}\right)$		
	NM*	MCST#	% Error	NM*	MCST#	% Error
0	2.6457	2.6457	0	6.1733	6.1732	0.0016
0.1	2.8221	2.8224	0.0106	5.9970	5.9970	0
0.2	2.9985	2.9988	0.0100	5.8206	5.8206	0
0.3	3.1749	3.1749	0	5.6442	5.6434	0.0142
0.4	3.3512	3.3514	0.0060	5.4678	5.4674	0.0073
0.5	3.5276	3.5282	0.0170	5.2914	5.2912	0.0038
0.6	3.7040	3.7041	0.0027	5.1151	5.1144	0.0137
0.7	3.8804	3.8807	0.0077	4.9387	4.9385	0.0040
0.8	4.0568	4.0568	0	4.7623	4.7622	0.0021
0.9	4.2331	4.2332	0.0024	4.5859	4.5859	0
1.0	4.4095	4.4095	0	4.4095	4.4095	0

Time taken for LB and UB by MCST: 67.201382 seconds
* Time taken for LB and UB by NM: 3.957473 + 3.895574 = 7.853047 seconds

(b) Clamped–Hinged (CH) Boundary Condition

α_1	LB of Critical Buckling Load $\left(\underline{P}_{cr}\right)$			UB of Critical Buckling Load $\left(\bar{P}_{cr}\right)$		
	GWRM*	MCST#	% Error	GWRM*	MCST#	% Error
0	5.0399	5.0403	0.0079	11.7597	11.7583	0.0119
0.1	5.3759	5.3762	0.0056	11.4237	11.4227	0.0088
0.2	5.7119	5.7122	0.0053	11.0877	11.0876	0.0009
0.3	6.0479	6.0481	0.0033	10.7517	10.7516	0.0009
0.4	6.3838	6.3843	0.0078	10.4157	10.4156	0.0010
0.5	6.7198	6.7199	0.0015	10.0798	10.0796	0.0020
0.6	7.0558	7.0559	0.0014	9.7438	9.7434	0.0041
0.7	7.3918	7.3919	0.0014	9.4078	9.4069	0.0096
0.8	7.7278	7.7279	0.0013	9.0718	9.0715	0.0033
0.9	8.0638	8.0639	0.0012	8.7358	8.7358	0
1.0	8.3998	8.3998	0	8.3998	8.3998	0

Time taken for LB and UB by MCST: 1021.220816 seconds
* Time taken for LB and UB by GWRM: 6.889991 + 6.766945 = 13.656936 seconds

---

(Continued)

Table 4.2 (*Continued*)
LB and UB of Critical Buckling Loads P_{cr} in nN, Considering Young's Modulus as an Uncertain Parameter with $d = 1\,\text{nm}$, $e_0a = 1\,\text{nm}$, and $L = 10\,\text{nm}$.

(c) Clamped–Clamped (CC) Boundary Condition

α_1	LB of Critical Buckling Load $\left(\underline{P_{cr}}\right)$			UB of Critical Buckling Load $\left(\bar{P}_{cr}\right)$		
	GWRM*	MCST#	% Error	GWRM*	MCST#	% Error
0	8.3825	8.3835	0.0119	19.5592	19.5584	0.0041
0.1	8.9414	8.9422	0.0089	19.0004	19.0003	0.0005
0.2	9.5002	9.5009	0.0074	18.4416	18.4405	0.0060
0.3	10.0590	10.0616	0.0258	17.8827	17.8813	0.0078
0.4	10.6179	10.6180	0.0009	17.3239	17.3236	0.0017
0.5	11.1767	11.1774	0.0063	16.7650	16.7649	0.0006
0.6	11.7355	11.7360	0.0043	16.2062	16.2061	0.0006
0.7	12.2944	12.2950	0.0049	15.6474	15.6470	0.0026
0.8	12.8532	12.8535	0.0023	15.0885	15.0884	0.0007
0.9	13.4120	13.4122	0.0015	14.5297	14.5296	0.0007
1.0	13.9709	13.9709	0	13.9709	13.9709	0

\# Time taken for LB and UB by MCST: 789.324982 seconds
* Time taken for LB and UB by GWRM: $6.618879 + 6.595867 = 13.214746$ seconds

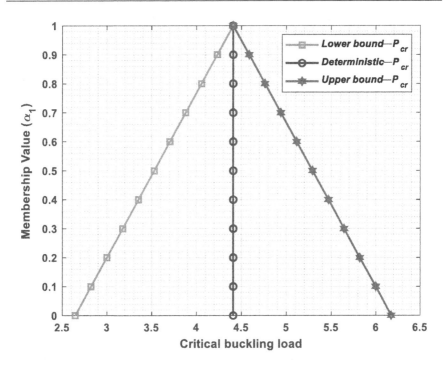

FIGURE 4.3   TFN of P_{cr} for HH boundary condition when E is an uncertain parameter.

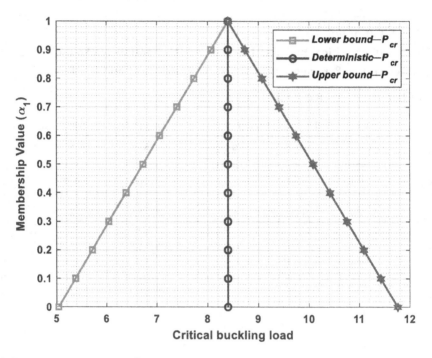

FIGURE 4.4   TFN of P_{cr} for CH boundary condition when E is an uncertain parameter.

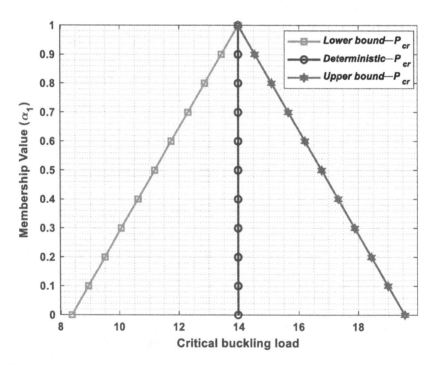

FIGURE 4.5   TFN of P_{cr} for CC boundary condition when E is an uncertain parameter.

bound. The tabular results reveal that the results obtained by the non-probabilistic methods and Monte Carlo Simulation Technique are quite agreeing with each other with very little errors. Also, it may be noted that the time taken by Navier's method is quite lesser than the Monte Carlo Simulation Technique. Figures 4.3–4.5 represent the uncertain critical buckling loads in the form Triangular Fuzzy Numbers for HH, CH, and CC boundary conditions, respectively. From these results, it is concluded that the spread of fuzziness or uncertainties in HH boundary is lower, whereas, in CC boundary condition, it is comparatively higher.

Table 4.3 and Figures 4.6–4.8 demonstrate the tabular and graphical results for the second case where the diameter of the nonlocal beam has been considered as uncertain considering the structural parameters as $\tilde{d} = (0.8, 1, 1.2)\,\text{nm}$, $E = 1\,TPa, e_0a = 1\,nm$, and $L = 10\,nm$. In the second case, the tabular results obtained by the non-probabilistic methods are also matching with the results obtained by Monte Carlo Simulation Technique. Also, the time taken by Navier's method is very less than that of the Monte Carlo Simulation Technique. Figures 4.6–4.8 represent the critical buckling loads in the form of Triangular Fuzzy Numbers for HH, CH, and CC boundary conditions, respectively. The propagation of uncertainties to the critical buckling loads with respect to the uncertain diameter is higher in case of clamped–lamped edge and lower in case of hinged–hinged edge.

Table 4.3
LB and UB of Critical Buckling Loads P_{cr} in nN, Considering Diameter as an Uncertain Parameter with $E = 1\,TPa$, $e_0a = 1\,nm$, and L = 10 nm.
(a) Hinged–Hinged (HH) Boundary Condition

α_2	LB of Critical Buckling Load $\left(\underline{P}_{cr}\right)$			UB of Critical Buckling Load $\left(\bar{P}_{cr}\right)$		
	NM*	MCST#	% Error	NM*	MCST#	% Error
0	1.8061	1.8067	0.0332	9.1436	9.1400	0.0394
0.1	1.9936	1.9952	0.0803	8.5491	8.5490	0.0012
0.2	2.1954	2.1955	0.0046	7.9841	7.9808	0.0413
0.3	2.4120	2.4124	0.0166	7.4475	7.4459	0.0215
0.4	2.6444	2.6444	0	6.9385	6.9383	0.0029
0.5	2.8931	2.8931	0	6.4560	6.4560	0
0.6	3.1590	3.1596	0.0190	5.9991	5.9981	0.0167
0.7	3.4427	3.4428	0.0029	5.5669	5.5667	0.0036
0.8	3.7452	3.7455	0.0080	5.1585	5.1584	0.0019
0.9	4.0672	4.0674	0.0049	4.7730	4.7730	0
1.0	4.4095	4.4095	0	4.4095	4.4095	0

Time taken for LB and UB by MCST: 51.188114 seconds
* Time taken for LB and UB by NM: 3.876123 + 3.921826 = 7.797949 seconds

Table 4.3 (*Continued*)

LB and UB of Critical Buckling Loads P_{cr} in nN, Considering Diameter as an Uncertain Parameter with $E = 1\,TPa$, $e_0 a = 1\,nm$, and $L = 10\,nm$.

(b) Clamped–Hinged (CH) boundary condition

α_2	LB of Critical Buckling Load $\left(\underline{P}_{cr}\right)$			UB of Critical Buckling Load $\left(\bar{P}_{cr}\right)$		
	GWRM*	MCST#	% Error	GWRM*	MCST#	% Error
0	3.4406	3.4417	0.0320	17.4178	17.4110	0.0390
0.1	3.7977	3.8007	0.0790	16.2853	16.2852	0.0006
0.2	4.1820	4.1822	0.0048	15.2090	15.2027	0.0414
0.3	4.5948	4.5955	0.0152	14.1869	14.1837	0.0226
0.4	5.0373	5.0374	0.0020	13.2172	13.2169	0.0023
0.5	5.5111	5.5112	0.0018	12.2981	12.2981	0
0.6	6.0176	6.0187	0.0183	11.4278	11.4259	0.0166
0.7	6.5581	6.5583	0.0030	10.6045	10.6042	0.0028
0.8	7.1343	7.1348	0.0070	9.8266	9.8264	0.0020
0.9	7.7477	7.7481	0.0052	9.0922	9.0921	0.0011
1.0	8.3998	8.3998	0	8.3998	8.3998	0

Time taken for LB and UB by MCST: 1028.984054 seconds
* Time taken for LB and UB by GWRM: 6.757198 + 6.849629 = 13.606827 seconds

(c) Clamped–Clamped (CC) Boundary Condition

α_1	LB of Critical Buckling Load $\left(\underline{P}_{cr}\right)$			UB of Critical Buckling Load $\left(\bar{P}_{cr}\right)$		
	GWRM*	MCST#	% Error	GWRM*	MCST#	% Error
0	5.7225	5.7227	0.0035	28.9700	28.9600	0.0345
0.1	6.3165	6.3166	0.0016	27.0864	27.0829	0.0129
0.2	6.9557	6.9558	0.0014	25.2962	25.2953	0.0036
0.3	7.6422	7.6425	0.0039	23.5962	23.5960	0.0008
0.4	8.3783	8.3794	0.0131	21.9834	21.9813	0.0096
0.5	9.1663	9.1664	0.0011	20.4548	20.4506	0.0205
0.6	10.0086	10.0092	0.0060	19.0072	19.0067	0.0026
0.7	10.9077	10.9095	0.0165	17.6379	17.6379	0
0.8	11.8661	11.8662	0.0008	16.3439	16.3439	0
0.9	12.8863	12.8865	0.0016	15.1225	15.1218	0.0046
1.0	13.9709	13.9709	0	13.9709	13.9709	0

Time taken for LB and UB by MCST: 792.197172 seconds
* Time taken for LB and UB by GWRM: 6.630419 + 6.642321 = 13.27274 seconds

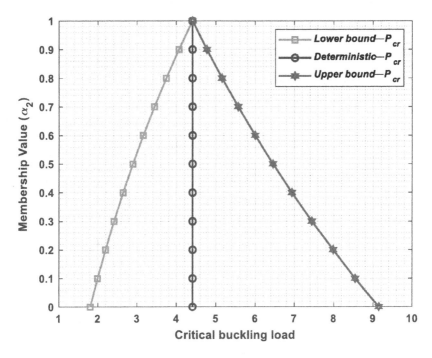

FIGURE 4.6   TFN of P_{cr} for HH boundary condition when d is an uncertain parameter.

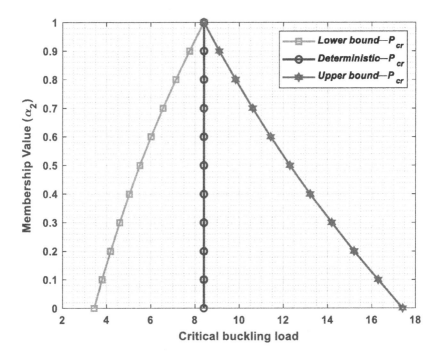

FIGURE 4.7   TFN of P_{cr} for CH boundary condition when d is an uncertain parameter.

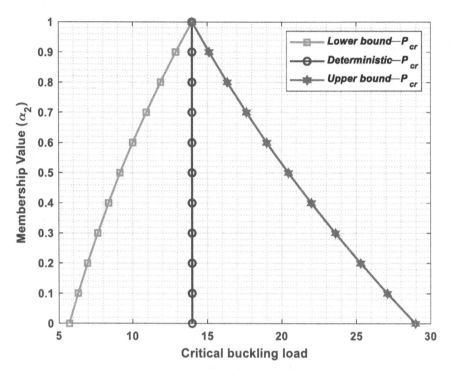

FIGURE 4.8 TFN of P_{cr} for CC boundary condition when d is an uncertain parameter.

Likewise, Table 4.4 and Figures 4.9–4.11 are the tabular and graphical results of critical buckling loads, respectively, where both Young's modulus and the diameter of the beam have been assumed as uncertain parameters with $\tilde{E} = (0.6, 1, 1.4)$ TPa, $\tilde{d} = (0.8, 1, 1.2)$ nm, $e_0 a = 1\,nm$, and $L = 10\,nm$. In this case, the tabular results are obtained by the non-probabilistic methods as the time taken to compute the critical buckling loads by Monte Carlo Simulation Technique is very high. Figures 4.9–4.11 are the uncertain critical buckling loads in the form Triangular Fuzzy Numbers for HH, CH, and CC boundary conditions, respectively.

The critical buckling loads for the uncertain systems in terms of Triangular Fuzzy Numbers are also compared for each boundary condition considering all the three cases in the form of graphical results which are given in Figures 4.12–4.14. Figure 4.12 represents the comparison for HH boundary condition whereas Figure 4.13 and Figure 4.14 demonstrate for CH and CC boundary conditions, respectively. From these figures, it can be observed that the spread of fuzziness or uncertainties in critical buckling load is comparatively less when Young's modulus is uncertain than that of uncertain diameter and uncertain Young's modulus and diameter. This is because uncertainties in Young's modulus and diameter are propagating to critical buckling load with a comparatively larger extent than that of uncertain Young's modulus and uncertain diameter. This trend is true for all the three boundary conditions.

Table 4.4

LB and UB of Critical Buckling Loads (P_{cr}) in nN, Considering Both Young's Modulus and Diameter as Uncertain Parameters with $e_0 a = 1$ nm and $L = 10$ nm

$\alpha_1 = \alpha_2$	LB of Critical Buckling Load $(\underline{P_{cr}})$			UB of Critical Buckling Load $(\overline{P_{cr}})$		
	HH	CH	CC	HH	CH	CC
0	1.0837	2.0643	3.4335	12.8010	24.3849	40.5580
0.1	1.2759	2.4305	4.0426	11.6268	22.1480	36.8375
0.2	1.4929	2.8438	4.7299	10.5390	20.0759	33.3910
0.3	1.7367	3.3082	5.5024	9.5328	18.1593	30.2032
0.4	2.0097	3.8284	6.3675	8.6037	16.3894	27.2595
0.5	2.3145	4.4089	7.3330	7.7472	14.7578	24.5457
0.6	2.6535	5.0547	8.4073	6.9590	13.2563	22.0484
0.7	3.0296	5.7712	9.5988	6.2350	11.8771	19.7545
0.8	3.4456	6.5636	10.9168	5.5712	10.6127	17.6515
0.9	3.9045	7.4378	12.3708	4.9639	9.4559	15.7274
1.0	4.4095	8.3998	13.9709	4.4095	8.3998	13.9709

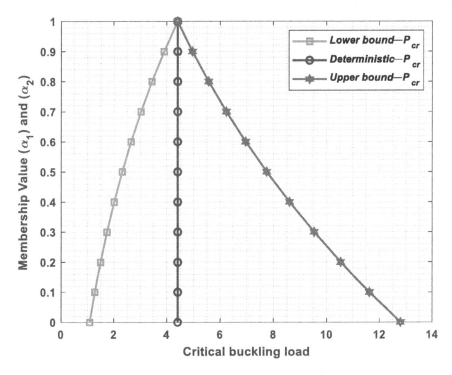

FIGURE 4.9 TFN of P_{cr} for HH boundary condition when E and d are uncertain parameters.

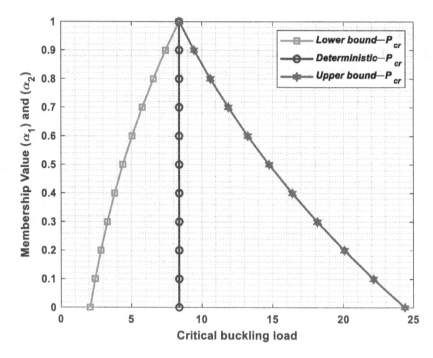

FIGURE 4.10   TFN of P_{cr} for CH boundary condition when E and d are uncertain parameters.

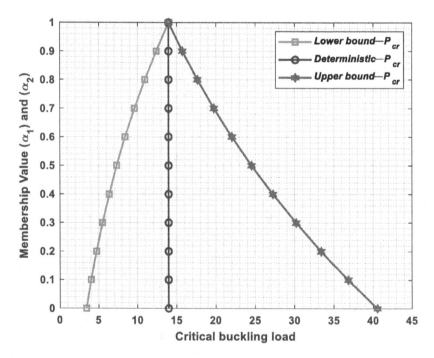

FIGURE 4.11   TFN of P_{cr} for CC boundary condition when E and d are uncertain parameters.

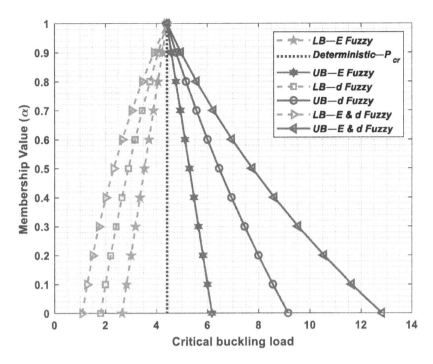

FIGURE 4.12   Comparisons of TFN of P_{cr} for HH boundary condition.

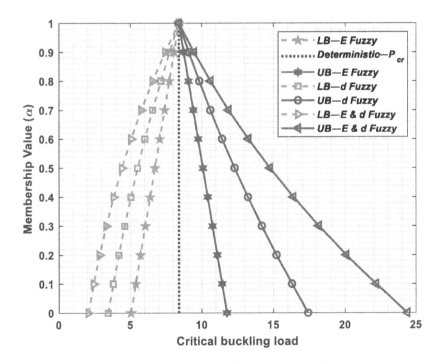

FIGURE 4.13   Comparisons of TFN of P_{cr} for CH boundary condition.

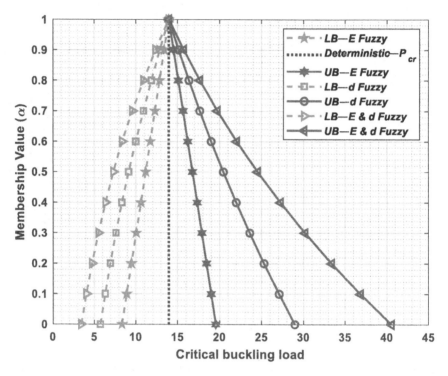

FIGURE 4.14 Comparisons of TFN of P_{cr} for CC boundary condition.

4.5 CONCLUDING REMARKS

This chapter addresses the modeling of an Euler–Bernoulli nanobeam under the influence of material uncertainties, focusing on uncertain parameters such as Young's modulus and the beam diameter. These material uncertainties are represented using fuzzy numbers, specifically Triangular Fuzzy Numbers, with the double parametric form adopted to effectively handle the fuzzy uncertainty. Two non-probabilistic methods, namely the double parametric form-based Navier's method (NM) and Galerkin Weighted Residual Method (GWRM), are introduced and applied to compute the critical buckling loads of the nonlocal beam. Notably, this approach also provides direct results for crisp and interval cases. Despite the necessity to develop numerical methods tailored to the double parametric form of fuzzy numbers, the results are presented in a general form that encapsulates the essence of uncertainties. The study considers three boundary conditions—hinged–hinged (HH), clamped–hinged (CH), and clamped–clamped (CC)—and employs a Monte Carlo Simulation Technique (MCST) for computing critical buckling loads in uncertain systems. The comparison of results obtained from non-probabilistic methods with deterministic models in special cases and uncertain models reveals robust agreement. Notably, non-probabilistic methods exhibit significantly reduced computation time compared to MCST, underscoring their efficiency. Furthermore, a parametric study demonstrates the propagation of uncertainties into the nonlocal system through critical buckling loads. It is observed that the spread of uncertainties is less pronounced when Young's modulus is uncertain, whereas greater uncertainties in both Young's modulus and diameter result in a higher spread of critical buckling loads, highlighting the impact of structural parameter uncertainties on system behavior.

BIBLIOGRAPHY

[1] Reddy JN (2007) Nonlocal theories for bending, buckling and vibration of beams. International Journal of Engineering Science 45: 288–307

[2] Eringen AC (1972) Nonlocal polar elastic continua. International Journal Engineering Science 10: 1–16

[3] Malikan M, Nguyen VB, Tornabene F (2018) Damped forced vibration analysis of single-walled carbon nanotubes resting on viscoelastic foundation in thermal environment using nonlocal strain gradient theory. Engineering Science and Technology, an International Journal 21: 778–786

[4] Malikan M (2019) On the buckling response of axially pressurized nanotubes based on a novel nonlocal beam theory. Journal of Applied and Computational Mechanics 5: 103–112

[5] Malikan M, Dimitri R, Tornabene F (2019) Transient response of oscillated carbon nanotubes with an internal and external damping. Composites Part B: Engineering 158: 198–205

[6] Sobhy M (2015) Thermoelastic response of FGM plates with temperature-dependent properties resting on variable elastic foundations. International Journal of Applied Mechanics 7(06): 1550082

[7] Zhu J, Lv Z, Liu H (2019) Thermo-electro-mechanical vibration analysis of nonlocal piezoelectric nanoplates involving material uncertainties. Composite Structures 208: 771–783

[8] Wang CM, Zhang YY, Ramesh SS, Kitipornchai S (2006) Buckling analysis of micro- and nano-rods/tubes based on nonlocal Timoshenko beam theory. Journal of Physics D: Applied Physics 39: 3904

5 Vibration of Nanobeam Based on One-Variable First-Order Shear Deformation Beam Theory in Uncertain Environment

5.1 INTRODUCTION

This chapter investigates uncertainty quantification in nanobeam vibration by integrating the one-variable first-order shear deformation beam theory with Eringen's nonlocal elasticity theory. Material uncertainties associated with mass density and Young's modulus are represented using Triangular Fuzzy Numbers. The governing equations for vibration are derived employing the von Kármán hypothesis and Hamilton's principle, leading to closed-form solutions for the simply supported boundary condition through Navier's method based on the double parametric form. A comparison of the obtained frequency parameters with those in previously published literature demonstrates a strong agreement in specific cases. Additionally, a Monte Carlo Simulation Technique (MCST) based on random sampling is utilized to comprehensively assess the natural frequencies of the nanobeam amid material uncertainties, providing insights into the variability of the system's response. In order to validate the results obtained through the uncertain model, the natural frequencies derived from Navier's method (NM) in terms of lower bound (LB) and upper bound (UB) are rigorously compared with those obtained through Monte Carlo Simulation Technique (MCST). This comparative analysis underscores the efficacy, accuracy, and robustness of the proposed uncertain model in capturing the dynamic behavior of the nanobeam under varying material conditions. The computation of lower and upper bounds of frequency parameters using the double parameter and graphical outputs demonstrating the sensitivity of the models are plotted on the basis of the Triangular Fuzzy Number. This uncertainty modeling and the delineation of frequency parameter bounds offer effective tools for engineering structure design and quality optimization in nanobeam applications.

DOI: 10.1201/9781003303107-5

5.2   MATHEMATICAL FORMULATION OF THE PROPOSED PROBLEM

The expression of displacement fields, according to one-variable first-order shear deformation beam theory, is given as [1–3]:

$$u_1\left(x,z,t\right) = u\left(x,t\right) - z\frac{\partial w\left(x,t\right)}{\partial x}$$

$$u_2\left(x,z,t\right) = 0 \tag{5.1}$$

$$u_3\left(x,z,t\right) = w\left(x,t\right) + B\frac{\partial^2 w}{\partial x^2}$$

Here, $u_1\left(x,z,t\right)$, $u_2\left(x,z,t\right)$, and $u_3\left(x,z,t\right)$ represent the displacement components in the x, y, and z directions, respectively. $u\left(x,t\right)$ and $w\left(x,t\right)$ represent the displacements of the neutral axis in the axial and transverse directions, respectively. $B = \dfrac{EI}{AG}$, where E is the Young's modulus, $I = \displaystyle\int_A z^2 \, dA$ is the moment of area, A is the area of cross-section, and G is the shear modulus.

The strain–displacement relations, according to the von Kármán hypothesis, are expressed as:

$$\begin{aligned}
\varepsilon_{xx} &= \frac{\partial u_1}{\partial x} + \frac{1}{2}\left(\frac{\partial u_3}{\partial x}\right)^2 \\
&= \frac{\partial u}{\partial x} - z\frac{\partial^2 w}{\partial x^2} + \frac{1}{2}\left(\frac{\partial w}{\partial x} + B\frac{\partial^3 w}{\partial x^3}\right)^2 \\
\gamma_{xz} &= \frac{\partial u_1}{\partial z} + \frac{\partial u_3}{\partial x} \\
&= B\frac{\partial^3 w}{\partial x^3}
\end{aligned} \tag{5.2}$$

The virtual strain energy $\left(\delta U\right)$ may be written as [1]:

$$\delta U = \iiint_V \left(\sigma_{xx}\delta\varepsilon_{xx} + \sigma_{xz}\delta\gamma_{xz}\right)dV$$

$$= \int_0^L \left[\begin{array}{l} N_{xx}\dfrac{\partial \delta u}{\partial x} - M_{xx}\dfrac{\partial^2 \delta w}{\partial x^2} + Q_{xz}B\dfrac{\partial^3 \delta w}{\partial x^3} + \\[2mm] N_{xx}\left(B^2\dfrac{\partial^3 w}{\partial x^3}\dfrac{\partial^3 \delta w}{\partial x^3} + B\dfrac{\partial^3 w}{\partial x^3}\dfrac{\partial \delta w}{\partial x} + B\dfrac{\partial w}{\partial x}\dfrac{\partial^3 \delta w}{\partial x^3} + \dfrac{\partial w}{\partial x}\dfrac{\partial \delta w}{\partial x}\right) \end{array}\right] dx \tag{5.3}$$

$$= \int_0^L \left[\begin{array}{l} -\dfrac{\partial N_{xx}}{\partial x}\delta u + \dfrac{\partial^2 M_{xx}}{\partial x^2}\delta w - B\dfrac{\partial^3 Q_{xz}}{\partial x^3}\delta w - B^2\left(\dfrac{\partial^3}{\partial x^3}\left(N_{xx}\dfrac{\partial^3 w}{\partial x^3}\right)\right)\delta w \\[3mm] -B\left(\dfrac{\partial}{\partial x}\left(N_{xx}\dfrac{\partial^3 w}{\partial x^3}\right)\right)\delta w - B\left(\dfrac{\partial^3}{\partial x^3}\left(N_{xx}\dfrac{\partial w}{\partial x}\right)\right)\delta w - \dfrac{\partial}{\partial x}\left(N_{xx}\dfrac{\partial w}{\partial x}\right)\delta w \end{array}\right] dx$$

where $M_{xx} = \int\limits_{A} z\sigma_{xx} dA$, $N_{xx} = \int\limits_{A} \sigma_{xx} dA$, and $Q_{xz} = \int\limits_{A} \sigma_{xz} dA$ are the local stress resultants of the nanobeam.

Now the virtual kinetic energy (δT) can be computed as [1]:

$$\delta T = \int\limits_{0}^{L} \left[\begin{array}{l} I_0 \dfrac{\partial u}{\partial t}\dfrac{\partial \delta u}{\partial t} + I_2 \dfrac{\partial^2 w}{\partial x \partial t}\dfrac{\partial^2 \delta w}{\partial x \partial t} + I_0 \dfrac{\partial w}{\partial t}\dfrac{\partial \delta w}{\partial t} \\ + I_0 B \dfrac{\partial w}{\partial t}\dfrac{\partial^3 \delta w}{\partial x^2 \partial t} + I_0 B \dfrac{\partial^3 w}{\partial x^2 \partial t}\dfrac{\partial \delta w}{\partial t} + I_0 B^2 \dfrac{\partial^3 w}{\partial x^2 \partial t}\dfrac{\partial^3 \delta w}{\partial x^2 \partial t} \end{array} \right] dx$$

$$= \int\limits_{0}^{L} \left[-I_0 \dfrac{\partial^2 u}{\partial t^2}\delta u - \left(I_2 \dfrac{\partial^4 w}{\partial x^2 \partial t^2} + I_0 \dfrac{\partial^2 w}{\partial t^2} + 2I_0 B \dfrac{\partial^4 w}{\partial x^2 \partial t^2} + B^2 I_0 \dfrac{\partial^6 w}{\partial x^4 \partial t^2} \right)\delta w \right] dx \tag{5.4}$$

where $I_0 = \rho A$ and $I_2 = \rho I$ are called mass moments of inertia.

Utilizing Equations (5.3) and (5.4) within Hamilton's principle, $\delta \Pi = \int\limits_{0}^{t} \delta(T - U) dt$, leads to the derivation of the equations of motion as:

$$\frac{\partial N_{xx}}{\partial x} = I_0 \frac{\partial^2 u}{\partial t^2} \tag{5.5}$$

$$\left[\begin{array}{l} -\dfrac{\partial^2 M_{xx}}{\partial x^2} + B \dfrac{\partial^3 Q_{xz}}{\partial x^3} + B^2 \left[\dfrac{\partial^3}{\partial x^3}\left(N_{xx}\dfrac{\partial^3 w}{\partial x^3} \right) \right] + B \left[\dfrac{\partial}{\partial x}\left(N_{xx}\dfrac{\partial^3 w}{\partial x^3} \right) \right] \\ + B \left[\dfrac{\partial^3}{\partial x^3}\left(N_{xx}\dfrac{\partial w}{\partial x} \right) \right] + \dfrac{\partial}{\partial x}\left(N_{xx}\dfrac{\partial w}{\partial x} \right) - I_2 \dfrac{\partial^4 w}{\partial x^2 \partial t^2} \\ - I_0 \left(\dfrac{\partial^2 w}{\partial t^2} + 2B \dfrac{\partial^4 w}{\partial x^2 \partial t^2} + B^2 \dfrac{\partial^6 w}{\partial x^4 \partial t^2} \right) \end{array} \right] = 0 \tag{5.6}$$

The Eq. (5.6) is further simplified as:

$$\left[\begin{array}{l} -\dfrac{\partial^2 M_{xx}}{\partial x^2} + B \dfrac{\partial^3 Q_{xz}}{\partial x^3} + N_{xx}\left(B^2 \dfrac{\partial^6 w}{\partial x^6} + 2B \dfrac{\partial^4 w}{\partial x^4} + \dfrac{\partial^2 w}{\partial x^2} \right) \\ - I_2 \dfrac{\partial^4 w}{\partial x^2 \partial t^2} - I_0 \left(\dfrac{\partial^2 w}{\partial t^2} + 2B \dfrac{\partial^4 w}{\partial x^2 \partial t^2} + B^2 \dfrac{\partial^6 w}{\partial x^4 \partial t^2} \right) \end{array} \right] = 0 \tag{5.7}$$

The local stress resultants can be reformulated using the Hookean stress–strain elasticity relation as:

$$M_{xx} = -EI \frac{\partial^2 w}{\partial x^2}$$

$$Q_{xz} = AGB \frac{\partial^3 w}{\partial x^3} \tag{5.8}$$

From the Eringen's nonlocal elasticity theory [4], we have:

$$\left(1-\left(e_0 a\right)^2 \frac{\partial^2}{\partial x^2}\right)\sigma_{ij} = C_{ijkl}\varepsilon_{kl} \tag{5.9}$$

In which, σ_{ij}, ε_{kl}, and C_{ijkl} are stress tensor, strain tensor, and elastic constant, respectively.

Further, from Eq. (5.9), one can have:

$$\left(1-\left(e_0 a\right)^2 \frac{\partial^2}{\partial x^2}\right)\sigma_{xx} = E\varepsilon_{xx} \tag{5.10}$$

$$\left(1-\left(e_0 a\right)^2 \frac{\partial^2}{\partial x^2}\right)\sigma_{xz} = 2G\varepsilon_{xz} \tag{5.11}$$

Using Eq. (5.10) and Eq. (5.11), the nonlocal stress resultants can be expressed as:

$$\left(1-\left(e_0 a\right)^2 \frac{\partial^2}{\partial x^2}\right)M_{xx} = -EI\frac{\partial^2 w}{\partial x^2} \tag{5.12}$$

$$\left(1-\left(e_0 a\right)^2 \frac{\partial^2}{\partial x^2}\right)Q_{xz} = AGB\frac{\partial^3 w}{\partial x^3} \tag{5.13}$$

By implementing Eq. (5.12) and Eq. (5.13) into Eq. (5.7), the governing equation of motion can be formulated as:

$$\left(1-\left(e_0 a\right)^2 \frac{\partial^2}{\partial x^2}\right)\left[\begin{array}{l} -N_{xx}\left(B^2 \frac{\partial^6 w}{\partial x^6} + 2B\frac{\partial^4 w}{\partial x^4} + \frac{\partial^2 w}{\partial x^2}\right) + I_2\left(\frac{\partial^4 w}{\partial x^2 \partial t^2}\right) \\ + I_0\left(\frac{\partial^2 w}{\partial t^2} + 2B\frac{\partial^4 w}{\partial x^2 \partial t^2} + B^2\frac{\partial^6 w}{\partial x^4 \partial t^2}\right) \end{array}\right]$$
$$= EI\frac{\partial^4 w}{\partial x^4} + AGB^2\frac{\partial^6 w}{\partial x^6} \tag{5.14}$$

In linear free vibration, the in-plane force resultant $\left(N_{xx}\right)$ can be disregarded from Eq. (5.14), leading to the following formulation for the governing equation of motion:

$$\left(1-\left(e_0 a\right)^2 \frac{\partial^2}{\partial x^2}\right)\left[I_2\left(\frac{\partial^4 w}{\partial x^2 \partial t^2}\right) + I_0\left(\frac{\partial^2 w}{\partial t^2} + 2B\frac{\partial^4 w}{\partial x^2 \partial t^2} + B^2\frac{\partial^6 w}{\partial x^4 \partial t^2}\right)\right]$$
$$= EI\frac{\partial^4 w}{\partial x^4} + AGB^2\frac{\partial^6 w}{\partial x^6} \tag{5.15}$$

5.2.1 MODELING WITH MATERIAL UNCERTAINTIES

In this chapter, Young's modulus and mass density of the nanobeam are characterized as Triangular Fuzzy Numbers (TFN) to represent material uncertainties, while other parameters remain crisp. The reason to choose material uncertainties as Triangular Fuzzy Numbers is due to their simplicity, interpretability, and computational efficiency. Triangular Fuzzy Numbers offer a straightforward representation of uncertainty, with only three parameters—the lower bound, upper bound, and modal value—required to define the fuzzy set. This simplicity makes Triangular Fuzzy Numbers easy to understand and work with, facilitating their application in engineering analyses. One advantage of using Triangular Fuzzy Numbers is their intuitive interpretation. The triangular shape of the fuzzy set visually conveys the range of possible values for the uncertain parameter, with the modal value representing the most likely or typical value, and the lower and upper bounds delineating the extent of uncertainty. This visual representation aids engineers and decision-makers in understanding the implications of uncertainty on the system under consideration. Additionally, Triangular Fuzzy Numbers are computationally efficient compared to other fuzzy number representations such as trapezoidal or Gaussian fuzzy numbers. The mathematical operations involved in manipulating Triangular Fuzzy Numbers are relatively straightforward, leading to faster computations and reduced computational overhead. This efficiency is particularly beneficial when conducting sensitivity analyses, optimization studies, or Monte Carlo Simulations involving large datasets or complex models.

The Triangular Fuzzy Numbers (TFN) assigned to the uncertain parameters are depicted as:

$$\tilde{E} = TFN\left(E_1, E_2, E_3\right) \qquad \text{and} \qquad \tilde{\rho} = TFN\left(\rho_1, \rho_2, \rho_3\right) \tag{5.16}$$

Here, E_1 and ρ_1 are lower bounds; E_2 and ρ_2 are modal values; and E_3 and ρ_3 are upper bounds of the Young's modulus and mass density of the nanobeam, respectively. By integrating the concept delineated in prelimnaries for the double parametric form, the Triangular Fuzzy Numbers $\tilde{E} = TFN\left(E_1, E_2, E_3\right)$ and $\tilde{\rho} = TFN\left(\rho_1, \rho_2, \rho_3\right)$ can be expressed in double parametric form as:

$$\tilde{E}\left(\alpha_1, \beta_1\right) = \left(E_1 - E_3\right)\alpha_1\beta_1 + \left(E_3 - E_1\right)\beta_1 + \left(E_2 - E_1\right)\alpha_1 + E_1,$$
$$\alpha_1, \beta_1 \in \begin{bmatrix} 0 & 1 \end{bmatrix} \tag{5.17}$$

$$\tilde{\rho}\left(\alpha_2, \beta_2\right) = \left(\rho_1 - \rho_3\right)\alpha_2\beta_2 + \left(\rho_3 - \rho_1\right)\beta_2 + \left(\rho_2 - \rho_1\right)\alpha_2 + \rho_1,$$
$$\alpha_2, \beta_2 \in \begin{bmatrix} 0 & 1 \end{bmatrix} \tag{5.18}$$

Substituting Eq. (5.17) and Eq. (5.18) into Eq. (5.15), the governing equation can be reformulated in double parametric form as:

$$
\left(1-\left(e_0 a\right)^2 \frac{\partial^2}{\partial x^2}\right)\left(\tilde{\rho}\left(\alpha_2, \beta_2\right)\right)\left[I\left(\frac{\partial^4 w}{\partial x^2 \partial t^2}\right)+A\left(\begin{array}{c}\frac{\partial^2 w}{\partial t^2}+2\left(\frac{\tilde{E}\left(\alpha_1, \beta_1\right) \times I}{AG}\right)\left(\frac{\partial^4 w}{\partial x^2 \partial t^2}\right)\\+\left(\frac{\tilde{E}\left(\alpha_1, \beta_1\right) \times I}{AG}\right)^2\left(\frac{\partial^6 w}{\partial x^4 \partial t^2}\right)\end{array}\right)\right]
$$

$$
=\tilde{E}\left(\alpha_1, \beta_1\right) I \frac{\partial^4 w}{\partial x^4}+\left(\frac{\left(\tilde{E}\left(\alpha_1, \beta_1\right) \times I\right)^2}{AG}\right) \frac{\partial^6 w}{\partial x^6} \tag{5.19}
$$

5.3 SOLUTION PROCEDURES

In this section, two methodologies, namely Navier's method and the Monte Carlo Simulation Technique, have been employed to ascertain the natural frequencies of a nanobeam subject to material uncertainties. Navier's method was specifically applied to address the simply supported–simply supported (SS) boundary condition, utilizing a double parametric form of Triangular Fuzzy Number theory. Concurrently, the Monte Carlo Simulation Technique was also utilized by converting Triangular Fuzzy Number representation into the interval form using a single parametric form.

5.3.1 NAVIER'S METHOD

In this study, Navier's method has been employed to analyze the vibration characteristics analytically under simply supported–simply supported (SS) boundary conditions. The transverse displacement according to Navier's approach can be expressed as [1–3]:

$$
w(x, t)=\sum_{n=1}^{\infty} W_n \sin \left(\frac{n \pi}{L} x\right) e^{i \omega_n t} \tag{5.20}
$$

In which W_n and ω_n are the displacement and frequency of the beam, respectively.
Plugging Eq. (5.20) in Eq. (5.19), the frequency parameter $\left(\omega^2\right)$ may be expressed as

$$
\omega_n^2=\frac{\tilde{E}\left(\alpha_1, \beta_1\right) I\left(\frac{n \pi}{L}\right)^4-\left(\frac{\left(\tilde{E}\left(\alpha_1, \beta_1\right) \times I\right)^2}{AG}\right)\left(\frac{n \pi}{L}\right)^6}{\left(\tilde{\rho}\left(\alpha_2, \beta_2\right)\right)\left(\begin{array}{l}\left(I\left(\frac{n \pi}{L}\right)^2-A+2 A\left(\frac{\tilde{E}\left(\alpha_1, \beta_1\right) \times I}{AG}\right)\left(\frac{n \pi}{L}\right)^2-\left(\frac{\tilde{E}\left(\alpha_1, \beta_1\right) \times I}{AG}\right)^2 A\left(\frac{n \pi}{L}\right)^4\right)+\left(e_0 a\right)^2\\\left(I\left(\frac{n \pi}{L}\right)^4-A\left(\frac{n \pi}{L}\right)^2+2 A\left(\frac{\tilde{E}\left(\alpha_1, \beta_1\right) \times I}{AG}\right)\left(\frac{n \pi}{L}\right)^4-\left(\frac{\tilde{E}\left(\alpha_1, \beta_1\right) \times I}{AG}\right)^2 A\left(\frac{n \pi}{L}\right)^6\right)\end{array}\right)} \tag{5.21}
$$

In the above Eq. (5.21), the shear modulus $G = \dfrac{\tilde{E}(\alpha_1, \beta_1)}{2(1+v)}$.

5.3.2 MONTE CARLO SIMULATION TECHNIQUE

Monte Carlo Simulation Technique is a powerful computational method used in uncertainty quantification to assess the variability and uncertainty associated with complex systems such as the vibration characteristics of nanobeams. In the context of nanobeam vibration, this technique involves selecting input parameters, such as material properties, geometrical dimensions, boundary conditions, and environmental factors, which contribute to uncertainty. Probability distributions are assigned to these parameters based on available data or expert knowledge, and random samples are generated from these distributions. These samples are then used in numerical simulations to determine the vibration characteristics of the nanobeam for each set of input parameters. Statistical analysis is performed on the simulation results to quantify variability and uncertainty, including measures such as mean, and standard deviation. Sensitivity analysis helps identify the most influential parameters, while validation and verification ensure the accuracy of the simulation results. Monte Carlo Simulation provides valuable insights into the probabilistic behavior of nanobeam vibration and aids in making informed decisions in engineering design under uncertain conditions.

In this problem, material properties such as mass density and Young's modulus, which exhibit uncertainties, are described using Triangular Fuzzy Numbers (TFNs). To facilitate analysis, these fuzzy material properties are transformed into the interval forms utilizing a single parametric expression outlined in preliminaries via α cut technique as:

$$\tilde{E} = TFN(E_1, E_2, E_3) = \left[\underline{E}(\alpha_1)\ \overline{E}(\alpha_1)\right] = \left[(E_2 - E_1)\alpha_1 + E_1 - (E_3 - E_2)\alpha_1 + E_3\right],$$
$$\alpha_1 \in [0\ 1]$$

$$\tilde{\rho} = TFN(\rho_1, \rho_2, \rho_3) = \left[\underline{\rho}(\alpha_2)\ \overline{\rho}(\alpha_2)\right] = \left[(\rho_2 - \rho_1)\alpha_2 + \rho_1 - (\rho_3 - \rho_2)\alpha_2 + \rho_3\right],$$
$$\alpha_2 \in [0\ 1]$$

Subsequently, 10,000 random points are generated from these intervals, representing uncertain parameters. These random points are then employed in the deterministic analysis of natural frequencies. Through repeated deterministic analyses for all random points, lower and upper bounds as well as the mean of natural frequencies are computed. The mean of the natural frequencies, obtained via the Monte Carlo Simulation Technique, serves as a representation of the results derived from the deterministic model.

5.4 NUMERICAL RESULTS AND DISCUSSION

The natural frequencies $\left(\omega_n\right)$ are calculated from Eq. (5.21) by implementing Navier's method. The modal values with their spreads, i.e., Triangular Tuzzy Numbers associated with uncertain parameters, are presented in Table 5.1, and graphical presentation of uncertain parameters is also illustrated in Figure 5.1. For different values of α, different intervals are generated, and for each interval, lower bound (LB) and upper bound (UB) of natural frequencies are determined by incorporating the double parametric form in the Navier's approach. The intervals that are

Table 5.1

Assigning Triangular Fuzzy Number (TFN) to Material Properties

Parameters	Modal Value	Triangular Fuzzy Number
Young's Modulus $\left(E\right)$	$E = 1\,TPa$	$\tilde{E} = TFN\left(0.9, 1.0, 1.1\right)TPa$
Mass Density $\left(\rho\right)$	$\rho = 2.24\,g\,/\,cm^3$	$\tilde{\rho} = TFN\left(2.0, 2.24, 2.48\right)g\,/\,cm^3$

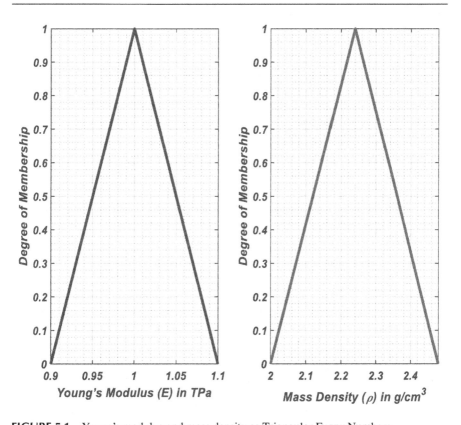

FIGURE 5.1 Young's modulus and mass density as Triangular Fuzzy Numbers.

Table 5.2

Intervals of Young's Modulus and Mass density for Different Degree of Memberships

α	$\underline{E}(\alpha)$	$\bar{E}(\alpha)$	$\underline{\rho}(\alpha)$	$\bar{\rho}(\alpha)$
0	0.9000	1.1000	2.0000	2.4800
0.1	0.9100	1.0900	2.0240	2.4560
0.2	0.9200	1.0800	2.0480	2.4320
0.3	0.9300	1.0700	2.0720	2.4080
0.4	0.9400	1.0600	2.0960	2.3840
0.5	0.9500	1.0500	2.1200	2.3600
0.6	0.9600	1.0400	2.1440	2.3360
0.7	0.9700	1.0300	2.1680	2.3120
0.8	0.9800	1.0200	2.1920	2.2880
0.9	0.9900	1.0100	2.2160	2.2640
1	1.0000	1.0000	2.2400	2.2400

considered in this research for the investigation are demonstrated in Table 5.2. In this investigation, a Single-Walled Carbon Nanotube [5] has been taken into consideration for the case study with Young's modulus $(\tilde{E}) = TFN(0.9,1,1.1)TPa$, mass density $(\tilde{\rho}) = TFN(2.0,2.24,2.48)g/cm^3$, Poisson's ratio $(v) = 0.28$, diameter $(d) = 1.1nm$, effective thickness $(h) = 0.342\,nm$, and, unless mentioned, $L = 10\,nm$. The frequency parameters $\left(\lambda = \omega L^2 \sqrt{\rho A/EI}\right)$ are validated by comparing with [35] in special cases for simply supported–simply supported boundary condition. For the computational purpose, $E = 1TPa$, Poisson's ratio $(v) = 0.18$, and diameter $(d) = 1nm$ have been taken. To demonstrate the accuracy of the frequency parameters $\left(\sqrt{\lambda}\right)$, the current findings are compared with those in [6], where the fuzziness in Young's modulus and mass density is neglected, along with the shear modulus and second moment of area. This comparison is presented in Table 5.3, showing an excellent agreement between the two sets of results for special case. The random sampling technique-based Monte Carlo Simulation Technique is also used to compute and verify the results of the present model. The tabular results for all the four modes of natural frequencies have been computed and demonstrated in Table 5.4(a–d) considering $\tilde{E} = (0.9,1,1.1)TPa$, $\rho = 2.24\,g/cm^3$, $e_0a = 0.1nm$, and $L = 10\,nm$ for SS nanobeam. The four modes of natural frequencies for the uncertain system are being computed through lower and upper bounds. The tabular results reveal that the results obtained by Navier's method and Monte Carlo Simulation Technique are well agreeing with each other, demonstrating the effectiveness of the approach.

Table 5.3

Frequency Parameters for SS Nanobeam with [6] for Special Cases

$\dfrac{e_0 a}{L}$	$\sqrt{\lambda_1}$		$\sqrt{\lambda_2}$		$\sqrt{\lambda_3}$		$\sqrt{\lambda_4}$	
	Present	[6]	Present	[6]	Present	[6]	Present	[6]
0	3.1416	3.1416	6.2832	6.2832	9.4248	9.4248	12.5662	12.5662
0.1	3.0685	3.0685	5.7817	5.7817	8.0400	8.0400	9.9161	9.9161
0.3	2.6800	2.6800	4.3013	4.3013	5.4422	5.4422	6.3630	6.3630
0.5	2.3022	2.3022	3.4604	3.4604	4.2941	4.2941	4.9820	4.9820

Table 5.4

Comparison of Natural Frequencies with $E = (0.9, 1, 1.1)\,TPa$, $\rho = 2.24\,g/cm^3$, and $e_0 a = 0.1\,nm$ for SS Nanobeam

(a) Fundamental Natural Frequencies in THz

α_1	Lower Bound of Fundamental Natural Frequencies $(\underline{\omega}_1)$		Upper Bound of Fundamental Natural Frequencies $(\bar{\omega}_1)$	
	NM	MCST	NM	MCST
0	0.0940	0.0940	0.1039	0.1039
0.1	0.0945	0.0945	0.1034	0.1034
0.2	0.0950	0.0950	0.1029	0.1029
0.3	0.0955	0.0955	0.1025	0.1025
0.4	0.0960	0.0960	0.1020	0.1020
0.5	0.0965	0.0965	0.1015	0.1015
0.6	0.0971	0.0971	0.1010	0.1010
0.7	0.0976	0.0976	0.1005	0.1005
0.8	0.0981	0.0981	0.1000	0.1000
0.9	0.0986	0.0986	0.0996	0.0995
1.0	0.0991	0.0991	0.0991	0.0991

(b) Second-Mode Natural Frequencies in THz

α_1	Lower Bound of Second-Mode Natural Frequencies $(\underline{\omega}_2)$		Upper Bound of Second-Mode Natural Frequencies $(\bar{\omega}_2)$	
	NM	MCST	NM	MCST
0	0.3950	0.3950	0.4367	0.4367
0.1	0.3972	0.3972	0.4347	0.4347
0.2	0.3994	0.3994	0.4327	0.4327
0.3	0.4015	0.4016	0.4307	0.4307
0.4	0.4037	0.4037	0.4287	0.4287
0.5	0.4058	0.4058	0.4267	0.4267
0.6	0.4080	0.4080	0.4246	0.4246
0.7	0.4101	0.4101	0.4226	0.4226
0.8	0.4122	0.4122	0.4205	0.4205
0.9	0.4143	0.4143	0.4185	0.4185
1.0	0.4164	0.4164	0.4164	0.4164

Table 5.4 (*Continued*)
Comparison of Natural Frequencies with $E = (0.9, 1, 1.1)\,TPa$, $\rho = 2.24\,g/cm^3$, and $e_0 a = 0.1\,nm$ for SS Nanobeam

(c) Third-Mode Natural Frequencies in THz

α_1	Lower Bound of Third-Mode Natural Frequencies $(\underline{\omega}_3)$		Upper Bound of Third-Mode Natural Frequencies $(\overline{\omega}_3)$	
	NM	MCST	NM	MCST
0	0.9859	0.9859	1.0899	1.0899
0.1	0.9913	0.9913	1.0850	1.0850
0.2	0.9968	0.9968	1.0800	1.0800
0.3	1.0022	1.0022	1.0750	1.0750
0.4	1.0075	1.0075	1.0699	1.0699
0.5	1.0129	1.0129	1.0649	1.0649
0.6	1.0182	1.0182	1.0598	1.0598
0.7	1.0235	1.0235	1.0547	1.0547
0.8	1.0288	1.0288	1.0495	1.0495
0.9	1.0340	1.0340	1.0444	1.0444
1.0	1.0392	1.0392	1.0392	1.0392

(d) Fourth-Mode Natural Frequencies in THz

α_1	Lower Bound of Fourth-Mode Natural Frequencies $(\underline{\omega}_4)$		Upper Bound of Third-Mode Natural Frequencies $(\overline{\omega}_4)$	
	NM	MCST	NM	MCST
0	2.2136	2.2137	2.4473	2.4473
0.1	2.2259	2.2259	2.4361	2.4361
0.2	2.2381	2.2382	2.4249	2.4249
0.3	2.2502	2.2503	2.4137	2.4136
0.4	2.2623	2.2623	2.4024	2.4023
0.5	2.2743	2.2743	2.3910	2.3910
0.6	2.2862	2.2862	2.3796	2.3796
0.7	2.2981	2.2981	2.3681	2.3681
0.8	2.3099	2.3099	2.3566	2.3566
0.9	2.3217	2.3217	2.3450	2.3450
1.0	2.3334	2.3334	2.3334	2.3334

To investigate the propagation of uncertainty in frequency parameters, three cases are examined. In the first case, Young's modulus $(\tilde{E}) = (0.9, 1, 1.1)\,TPa$ is considered as fuzzy or imprecise. In the second case, mass density $(\tilde{\rho}) = (2.0, 2.24, 2.48)\,g/cm^3$ is treated as imprecise. In the third case, both Young's modulus and mass density are considered as fuzz or imprecise. Table 5.5 illustrates the uncertainty in natural frequencies for both the lower and upper bounds of the simply supported (SS) boundary conditions for the first four modes of vibration of the nanobeam. The analysis is conducted with $E = (0.\tilde{9}, 1, 1.1)\,TPa$, $\rho = 2.24\,g/cm^3$, and $e_0 a = 0.5\,nm$ while keeping

Table 5.5

Natural Frequency (ω) in THz with $E = (0.9, 1, 1.0)\,TPa$, $\rho = 2.24\,g/cm^3$, and $e_0 a = 0.5\,nm$ for SS Nanobeam

α_2	$\beta_1 = 0$ (Lower Bound)				$\beta_2 = 1$ (Upper Bound)			
	$\underline{\omega}_1$	$\underline{\omega}_2$	$\underline{\omega}_3$	$\underline{\omega}_4$	$\bar{\omega}_1$	$\bar{\omega}_2$	$\bar{\omega}_3$	$\bar{\omega}_4$
0	0.0929	0.3776	0.8958	1.8891	0.1027	0.4175	0.9903	2.0885
0.1	0.0934	0.3797	0.9007	1.8996	0.1022	0.4156	0.9858	2.0790
0.2	0.0939	0.3818	0.9057	1.9100	0.1017	0.4136	0.9813	2.0694
0.3	0.0944	0.3838	0.9106	1.9203	0.1013	0.4117	0.9767	2.0598
0.4	0.0949	0.3859	0.9155	1.9306	0.1008	0.4098	0.9721	2.0502
0.5	0.0954	0.3879	0.9203	1.9409	0.1003	0.4079	0.9675	2.0405
0.6	0.0959	0.3900	0.9251	1.9511	0.0998	0.4059	0.9629	2.0307
0.7	0.0964	0.3920	0.9300	1.9612	0.0994	0.4040	0.9583	2.0209
0.8	0.0969	0.3940	0.9347	1.9713	0.0989	0.4020	0.9536	2.0111
0.9	0.0974	0.3960	0.9395	1.9813	0.0984	0.4000	0.9489	2.0012
1	0.0979	0.3980	0.9442	1.9913	0.0979	0.3980	0.9442	1.9913

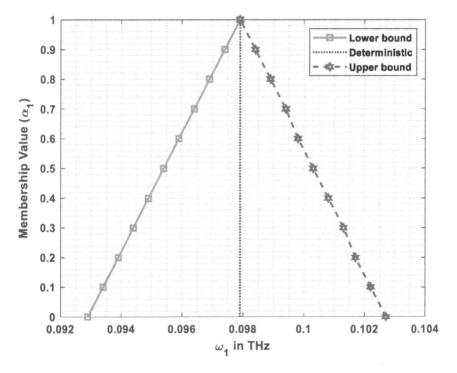

FIGURE 5.2 Fundamental natural frequencies in the form of TFN for SS boundary condition with $\tilde{E} = (0.9, 1, 1.1)\,TPa$.

all other parameters same as previous. Additionally, Figure 5.2 visualizes the propagation of uncertainty in the fundamental natural frequency when the Young's modulus is fuzzy. Similarly, Table 5.6 illustrates the propagation of uncertainty in the first four natural frequencies, displaying both lower and upper bounds for the second case. Furthermore, Figure 5.3 visually presents the fuzzy output of the fundamental

Table 5.6

Natural Frequency (ω) in THz with $E = 1\,TPa$, $\tilde{\rho} = (2.0, 2.24, 2.48)\,g/cm^3$, and $e_0 a = 0.5\,nm$ for SS Nanobeam

α_2	$\beta_1 = 0$ (Lower Bound)				$\beta_2 = 1$ (Upper Bound)			
	$\underline{\omega}_1$	$\underline{\omega}_2$	$\underline{\omega}_3$	$\underline{\omega}_4$	$\bar{\omega}_1$	$\bar{\omega}_2$	$\bar{\omega}_3$	$\bar{\omega}_4$
0	0.1036	0.4212	0.9993	2.1074	0.0930	0.3783	0.8974	1.8925
0.1	0.1030	0.4187	0.9933	2.0949	0.0935	0.3801	0.9017	1.9017
0.2	0.1024	0.4163	0.9875	2.0825	0.0940	0.3820	0.9062	1.9111
0.3	0.1018	0.4138	0.9818	2.0705	0.0944	0.3839	0.9107	1.9206
0.4	0.1012	0.4115	0.9761	2.0586	0.0949	0.3858	0.9153	1.9302
0.5	0.1006	0.4091	0.9706	2.0469	0.0954	0.3878	0.9199	1.9400
0.6	0.1001	0.4068	0.9651	2.0354	0.0959	0.3898	0.9246	1.9499
0.7	0.0995	0.4046	0.9598	2.0241	0.0964	0.3918	0.9294	1.9600
0.8	0.0990	0.4024	0.9545	2.0130	0.0969	0.3938	0.9343	1.9703
0.9	0.0984	0.4002	0.9493	2.0021	0.0974	0.3959	0.9392	1.9807
1	0.0979	0.3980	0.9442	1.9913	0.0979	0.3980	0.9442	1.9913

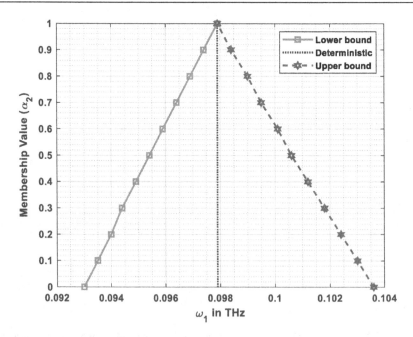

FIGURE 5.3 Fundamental natural frequencies in the form of TFN for SS boundary condition with $\tilde{\rho} = (2.0, 2.24, 2.48)\,g/cm^3$.

natural frequency for the second case with $E = 1\,TPa$, $\tilde{\rho} = (2.0, 2.24, 2.48)\,g/cm^3$, and $e_0 a = 0.5\,nm$. In the third case, both Young's modulus and mass density are considered fuzzy, and the outputs are presented in terms of fuzziness for various modes. This is depicted in Table 5.7 and Figure 5.4 for clarity and comparison.

Table 5.7
Natural Frequency (ω) in THz with $\tilde{E} = (0.9, 1, 1.1)\,TPa$, $\tilde{\rho} = (2.0, 2.24, 2.48)\,g/cm^3$, and $e_0 a = 0.5\,nm$ for SS Nanobeam

$\alpha_1 = \alpha_2$	$\beta_1 = \beta_2 = 0$ (Lower Bound)				$\beta_1 = \beta_2 = 1$ (Upper Bound)			
	$\underline{\omega}_1$	$\underline{\omega}_2$	$\underline{\omega}_3$	$\underline{\omega}_4$	$\overline{\omega}_1$	$\overline{\omega}_2$	$\overline{\omega}_3$	$\overline{\omega}_4$
0	0.0983	0.3996	0.9480	1.9992	0.0976	0.3967	0.9412	1.9849
0.1	0.0983	0.3994	0.9476	1.9984	0.0976	0.3969	0.9415	1.9854
0.2	0.0982	0.3993	0.9472	1.9975	0.0976	0.3970	0.9417	1.9860
0.3	0.0982	0.3991	0.9468	1.9967	0.0977	0.3971	0.9420	1.9867
0.4	0.0981	0.3989	0.9464	1.9959	0.0977	0.3972	0.9423	1.9873
0.5	0.0981	0.3988	0.9460	1.9950	0.0977	0.3974	0.9426	1.9879
0.6	0.0981	0.3986	0.9456	1.9943	0.0978	0.3975	0.9429	1.9886
0.7	0.0980	0.3985	0.9453	1.9935	0.0978	0.3976	0.9432	1.9892
0.8	0.0980	0.3983	0.9449	1.9927	0.0978	0.3977	0.9436	1.9899
0.9	0.0979	0.3982	0.9446	1.9920	0.0979	0.3979	0.9439	1.9906
1	0.0979	0.3980	0.9442	1.9913	0.0979	0.3980	0.9442	1.9913

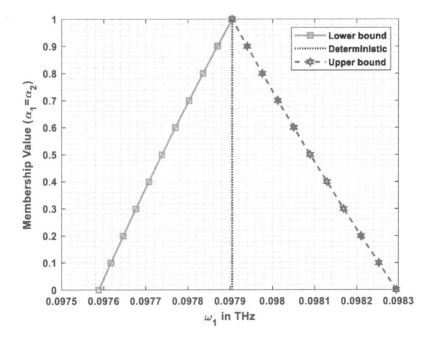

FIGURE 5.4 Fundamental natural frequencies in the form of TFN for SS boundary condition with $\tilde{E} = (0.9, 1, 1.1)\,TPa$ and $\tilde{\rho} = (2.0, 2.24, 2.48)\,g/cm^3$.

Upon analyzing the graphical and tabular outcomes, it becomes apparent that opting for $\alpha = 0$ yields frequency parameters represented in interval forms, encompassing both lower and upper bounds. Conversely, selecting $\alpha = 1$ results in deterministic (or exact or crisp) frequency parameters. Notably, $\beta = 0$ represents the lower bound, while $\beta = 1$ represents the upper bound of the frequency parameters. Interestingly, when $\alpha = 1$, both the upper and lower bounds converge to identical results, aligning with the deterministic values. In all the aforementioned cases, the computations are conducted while considering the nonlocal parameter $e_0 a = 0.5\,nm$.

Moreover, all uncertain frequencies for the aforementioned three cases are compared, as depicted in Figure 5.5. This figure illustrates the fundamental uncertain natural frequency for the SS boundary condition. As depicted in Figure 5.5, a larger width or fuzziness is indeed evident for both the first and second cases when uncertainties are associated with Young's modulus and mass density separately. Conversely, lower width is apparent in the figures when both mass density and Young's modulus are treated as fuzzy. This discrepancy in width or fuzziness can be attributed to the intricate interplay of uncertainties within the system. When uncertainties are considered separately for Young's modulus and mass density, each parameter contributes its own variability to the overall uncertainty in the natural frequency. Consequently, a broader range of possible values emerge, reflecting a larger width or fuzziness. Conversely, treating both mass density and Young's modulus as fuzzy variables simultaneously introduces the potential for their uncertainties to interact in complex ways. It is plausible that the combined effect of these uncertainties results in a more constrained range of possible values for the natural frequency. This constraint may arise if the uncertainties in one parameter offset or counterbalance those in the other, leading to a more focused or narrower distribution of possible

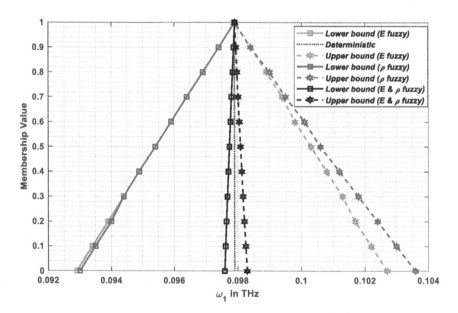

FIGURE 5.5 Comparison of fundamental frequencies for all the cases for SS boundary condition.

outcomes. In essence, the difference in width or fuzziness underscores the intricate dynamics of individual and combined uncertainties within the system, emphasizing the necessity of comprehensive consideration in uncertainty quantification studies. Understanding these implications is pivotal for informed decision-making in engineering design and optimization endeavors, as it provides insights into the nuanced effects of different uncertainty sources on system behavior.

5.5 CONCLUDING REMARKS

In this chapter, the propagation of uncertainty in nanobeam vibration has been explored using a combined framework of Eringen's nonlocal elasticity theory and the one-variable first-order shear deformation beam theory. Material uncertainties concerning mass density and Young's modulus in the nanobeam model have been represented as Triangular Fuzzy Numbers. The governing vibration equations were derived employing the von Kármán hypothesis and Hamilton's principle, and closed-form solutions for the simply supported boundary conditions were obtained through the double parametric form-based Navier's method. To authenticate the outcomes derived from the uncertain model, the natural frequencies determined using Navier's method (NM) in terms of lower bound (LB) and upper bound (UB) are meticulously contrasted with those acquired through Monte Carlo Simulation Technique (MCST). This comparative examination emphasizes the efficacy, accuracy, and robustness of the proposed uncertain model. The study reveals that when uncertainty is attributed to Young's modulus and mass density separately in the first and second cases, respectively, a broader range or greater fuzziness is observed. Conversely, a narrower range or decreased fuzziness is evident in the figures when both mass density and Young's modulus are concurrently treated as fuzzy variables. This disparity underscores the influence of individual versus combined uncertainties on the overall fuzziness of the natural frequency.

BIBLIOGRAPHY

[1] Jena, S.K., Chakraverty, S. and Malikan, M., 2020. Vibration and buckling characteristics of nonlocal beam placed in a magnetic field embedded in Winkler—Pasternak elastic foundation using a new refined beam theory: An analytical approach. *The European Physical Journal Plus*, *135*(2), pp.1–18.

[2] Malikan, M., Nguyen, V.B. and Tornabene, F., 2018. Damped forced vibration analysis of single-walled carbon nanotubes resting on viscoelastic foundation in thermal environment using nonlocal strain gradient theory. *Engineering Science and Technology, an International Journal*, *21*(4), pp. 778–786.

[3] Malikan, M., Dimitri, R. and Tornabene, F., 2019. Transient response of oscillated carbon nanotubes with an internal and external damping. *Composites Part B: Engineering*, *158*, pp. 198–205.

[4] Eringen, A.C., 1972. Nonlocal polar elastic continua. *International Journal of Engineering Science*, *10*(1), pp. 1–16.

[5] Zhen, Y.X., Wen, S.L. and Tang, Y., 2019. Free vibration analysis of viscoelastic nanotubes under longitudinal magnetic field based on nonlocal strain gradient Timoshenko beam model. *Physica E: Low-Dimensional Systems and Nanostructures*, *105*, pp. 116–124.

[6] Wang, C.M., Zhang, Y.Y. and He, X.Q., 2007. Vibration of nonlocal Timoshenko beams. *Nanotechnology*, *18*(10), p. 105401.

6 Stability Analysis of Timoshenko Nanobeam with Uncertainty

6.1 INTRODUCTION

This chapter aims to analyze the effects of material uncertainties on the stability of Timoshenko nanobeam. Here, the uncertainties are considered to be associated with the diameter, length, and Young's modulus of the nanobeam in terms of a special fuzzy number, namely Symmetric Gaussian Fuzzy Number (SGFN). The governing equations for the stability analysis of the uncertain model are obtained by incorporating the Timoshenko beam theory along with Hamilton's principle and the double parametric form of fuzzy numbers. The small-scale effect of the nanobeam is addressed by Eringen's elasticity theory, and the double parametric form-based Navier's method is employed to calculate the results of the uncertain models for the lower bound (LB) and upper bound (UB) of the buckling loads. The results obtained by the uncertain model are validated with the deterministic model in a particular case, as well as with the results obtained by Monte Carlo Simulation (MCS) technique in terms of the lower bound (LB) and upper bound (UB) of the buckling loads, exhibiting a robust agreement. Further, a parametric study is conducted to investigate the fuzziness or spreads of the buckling loads with respect to different uncertain parameters.

6.2 FORMULATION OF THE PROPOSED MODEL

The displacement field, as per Timoshenko beam theory, is stated as [1–2]:

$$
\begin{aligned}
u_1(x,z,t) &= u(x,t) + z\varphi(x,t) \\
u_2(x,z,t) &= 0 \\
u_3(x,z,t) &= w(x,t)
\end{aligned}
\tag{6.1}
$$

In which $u(x,t)$ is the axial displacement, $w(x,t)$ is the transverse displacement, and $\varphi(x,t)$ denotes the rotation of cross-section.

Using strain–displacement relationship, the Lagrangian strain may be obtained as:

$$
\begin{aligned}
\varepsilon_{xx} &= \frac{\partial u}{\partial x} + z\frac{\partial \varphi}{\partial x} \\
\gamma_{xz} &= \varphi + \frac{\partial w}{\partial x}
\end{aligned}
\tag{6.2}
$$

DOI: 10.1201/9781003303107-6

Here, ε_{xx} represents the normal strain, and γ_{xz} denotes the transverse shear strain. The strain energy U of the system is given by:

$$U = \int_0^L \int_A \left(\sigma_{xx}\varepsilon_{xx} + \sigma_{xz}\gamma_{xz}\right) dA dx \tag{6.3}$$

Taking variation, the virtual strain energy δU is presented as:

$$\delta U = \int_0^L \int_A \left(\sigma_{xx}\delta\varepsilon_{xx} + \sigma_{xz}\delta\gamma_{xz}\right) dA dx$$

$$= \int_0^L \int_A \left[\sigma_{xx}\delta\left(\frac{\partial u}{\partial x} + z\frac{\partial \varphi}{\partial x}\right) + \sigma_{xz}\delta\left(\varphi + \frac{\partial w}{\partial x}\right)\right] dA dx \tag{6.4}$$

$$= \int_0^L \left[N_{xx}\frac{\partial\delta u}{\partial x} + M_{xx}\frac{\partial\delta\varphi}{\partial x} + Q\left(\delta\varphi + \frac{\partial\delta w}{\partial x}\right)\right] dx$$

In which, the stress resultants are given by $\left(N_{xx}, M_{xx}, Q\right) = \int_A \left(\sigma_{xx}, z\sigma_{xx}, \sigma_{xz}\right) dA$.

The variation of work done $\left(\delta W\right)$ by mechanical compressive loads is obtained as [3–4]:

$$\delta W = -\int_0^L \left[-N_{xx}^0\left(\frac{\partial^2 w}{\partial x^2}\right)\right]\delta w dx \tag{6.5}$$

where N_{xx}^0 is the axial compressive force. By applying the variational principle and collecting the coefficients of $\delta u, \delta w$, and $\delta\phi$, one will obtain [3]:

$$\frac{\partial N_{xx}}{\partial x} = 0$$

$$\frac{\partial Q}{\partial x} = N_{xx}^0 \frac{\partial^2 w}{\partial x^2} \tag{6.6}$$

$$\frac{\partial M_{xx}}{\partial x} = Q_{xx}$$

The Eringen's constitutive equation for the one-dimensional structure is stated as [5]:

$$\left(1 - \mu\frac{\partial^2}{\partial x^2}\right)\sigma_{xx} = E\varepsilon_{xx} \tag{6.7}$$

Where, $\sigma_{xx}, \varepsilon_{xx}$, E, and $\mu = (e_0 a)^2$ are normal stress, normal strain, Young's modulus, and nonlocal parameter, respectively. Likewise, the constitutive relation for the shear stress and strain is given as:

$$\sigma_{xz} = G\gamma_{xz} \tag{6.8}$$

In which σ_{xz}, γ_{xz}, and G are transverse shear stress, transverse shear strain, and shear modulus, respectively. Multiplying zdA in Eq. (6.7) and integrating Eq. (6.7) as well as Eq. (6.8) over the area A, one may have:

$$\left(1 - \mu \frac{\partial^2}{\partial x^2}\right) M_{xx} = EI \frac{\partial \phi}{\partial x} \tag{6.9}$$

$$Q_{xx} = k_s AG \left(\phi + \frac{\partial w}{\partial x}\right) \tag{6.10}$$

Where k_s is the shear correction factor, and $I = \int_A z^2 dA$ is the second moment of inertia.

6.2.1 MODELING WITH MATERIAL UNCERTAINTIES

In this chapter, material uncertainties are associated with the diameter (d), length (L), and Young's modulus (E) of the nanobeam as Symmetric Gaussian Fuzzy Number (SGFN), whereas other parameters such as shear correction factor (k_s), mass density (ρ), and Poisson's ratio (v) are restricted to crisp form only. The Symmetric Gaussian Fuzzy Number (SGFN) associated with the uncertain parameters is presented as:

$$\tilde{d} = sgfn(\overline{d}, \sigma_1, \tilde{\sigma}_1), \; \tilde{L} = sgfn(\overline{L}, \sigma_2, \sigma_2), \text{ and } \tilde{E} = \tilde{s}gfn(\overline{E}, \sigma_3, \sigma_3) \tag{6.11}$$

Using Eqs. (6.9–6.11) in Eq. (6.6) and putting $N_{xx}^0 = P$, where P is the applied compressive force due to mechanical load, the uncertain governing equations for the stability analysis of Timoshenko beam in terms of Symmetric Gaussian fuzzy parameters are given as:

$$\left(k_s \times \frac{\pi \tilde{d}^2}{4} \times \frac{\tilde{E}}{2(1+v)}\right)\left(\frac{\partial \phi}{\partial x} + \frac{\partial^2 w}{\partial x^2}\right) - P\left(\frac{\partial^2 w}{\partial x^2}\right) = 0 \tag{6.12}$$

$$\left(\tilde{E} \times \frac{\pi \tilde{d}^4}{64}\right)\left(\frac{\partial^2 \phi}{\partial x^2}\right) + \mu P\left(\frac{\partial^3 w}{\partial x^3}\right) - \left(k_s \times \frac{\pi \tilde{d}^2}{4} \times \frac{\tilde{E}}{2(1+v)}\right)\left(\phi + \frac{\partial w}{\partial x}\right) = 0 \tag{6.13}$$

Using single parametric form or $\alpha - cut$ approach as given in preliminaries, Eq. (6.11) can be converted into the interval forms as:

$$\tilde{d} = sgfn\left(\bar{d}, \sigma_1, \sigma_1\right) = \left[\bar{d} - \sqrt{-2\sigma_1^2 \ln(\alpha)}, \ \ \bar{d} + \sqrt{-2\sigma_1^2 \ln(\alpha)}\right]$$

$$\tilde{L} = sgfn\left(\bar{L}, \sigma_2, \sigma_2\right) = \left[\bar{L} - \sqrt{-2\sigma_2^2 \ln(\alpha)}, \ \ \bar{L} + \sqrt{-2\sigma_2^2 \ln(\alpha)}\right] \qquad (6.14)$$

$$\tilde{E} = sgfn\left(\bar{E}, \sigma_3, \sigma_3\right) = \left[\bar{E} - \sqrt{-2\sigma_3^2 \ln(\alpha)}, \ \ \bar{E} + \sqrt{-2\sigma_3^2 \ln(\alpha)}\right] \text{with } \alpha \in (0, \ 1]$$

Plugging Eq. (6.14) into Eq. (6.12) and Eq. (6.13),

$$\left(k_s \times \frac{\pi\left(\left[\bar{d} - \sqrt{-2\sigma_1^2 \ln(\alpha)}, \ \ \bar{d} + \sqrt{-2\sigma_1^2 \ln(\alpha)}\right]\right)^2}{4} \times \right. \\ \left. \frac{}{\left[\bar{E} - \sqrt{-2\sigma_3^2 \ln(\alpha)}, \ \ \bar{E} + \sqrt{-2\sigma_3^2 \ln(\alpha)}\right]}{2(1+v)}\right) \times \left(\frac{\partial \varphi}{\partial x} + \frac{\partial^2 w}{\partial x^2}\right) - P\left(\frac{\partial^2 w}{\partial x^2}\right) = 0 \qquad (6.15)$$

$$\left(\frac{\left[\bar{E} - \sqrt{-2\sigma_3^2 \ln(\alpha)}, \ \ \bar{E} + \sqrt{-2\sigma_3^2 \ln(\alpha)}\right] \times}{\pi\left(\left[\bar{d} - \sqrt{-2\sigma_1^2 \ln(\alpha)}, \ \ \bar{d} + \sqrt{-2\sigma_1^2 \ln(\alpha)}\right]\right)^4}{64}\right)\left(\frac{\partial^2 \varphi}{\partial x^2}\right) + \mu P\left(\frac{\partial^3 w}{\partial x^3}\right)$$

$$- \left(k_s \times \frac{\pi\left(\left[\bar{d} - \sqrt{-2\sigma_1^2 \ln(\alpha)}, \ \ \bar{d} + \sqrt{-2\sigma_1^2 \ln(\alpha)}\right]\right)^2}{4} \times \\ \frac{}{\left[\bar{E} - \sqrt{-2\sigma_3^2 \ln(\alpha)}, \ \ \bar{E} + \sqrt{-2\sigma_3^2 \ln(\alpha)}\right]}{2(1+v)}\right)\left(\varphi + \frac{\partial w}{\partial x}\right) = 0 \qquad (6.16)$$

Now, by using double parametric form as given in the preliminaries, Eq. (6.15) and Eq. (6.16) can be expressed as:

$$\left(k_s \times \frac{\pi\left(\left(2\beta\sqrt{-2\sigma_1^2 \ln(\alpha)}\right) + \bar{d} - \sqrt{-2\sigma_1^2 \ln(\alpha)}\right)^2}{4} \times \\ \frac{}{\left(\left(2\beta\sqrt{-2\sigma_3^2 \ln(\alpha)}\right) + \bar{E} - \sqrt{-2\sigma_3^2 \ln(\alpha)}\right)}{2(1+v)}\right)$$

$$\left(\frac{\partial \phi}{\partial x} + \frac{\partial^2 w}{\partial x^2}\right) - P\left(\frac{\partial^2 w}{\partial x^2}\right) = 0 \qquad (6.17)$$

$$\left(\begin{array}{l} \dfrac{\left[\left(2\beta\sqrt{-2\sigma_3^2\ln(\alpha)}\right)+\bar{E}-\sqrt{-2\sigma_3^2\ln(\alpha)}\right]\times}{\pi\left[\left(2\beta\sqrt{-2\sigma_1^2\ln(\alpha)}\right)+\bar{d}-\sqrt{-2\sigma_1^2\ln(\alpha)}\right]^4}\left(\dfrac{\partial^2\phi}{\partial x^2}\right)+\mu P\left(\dfrac{\partial^3 w}{\partial x^3}\right)}{64} \\ -\left(k_s\times\dfrac{\pi\left[\left(2\beta\sqrt{-2\sigma_1^2\ln(\alpha)}\right)+\bar{d}-\sqrt{-2\sigma_1^2\ln(\alpha)}\right]^2}{\dfrac{4}{\left[\left(2\beta\sqrt{-2\sigma_3^2\ln(\alpha)}\right)+\bar{E}-\sqrt{-2\sigma_3^2\ln(\alpha)}\right]}}\times\left(\phi+\dfrac{\partial w}{\partial x}\right)\right)=0 \\ \qquad\qquad\qquad\qquad 2(1+v) \end{array}\right)$$

(6.18)

6.3 ANALYTICAL SOLUTION TO THE PROPOSED MODEL

In this chapter, Navier's technique has been utilized to solve the uncertain stability equation of Timoshenko nanobeam for hinged–hinged (H-H) boundary condition in terms of Symmetric Gaussian Fuzzy Number (SGFN). The transverse displacement $w(x,t)$ and the rotation of cross-section $\phi(x,t)$ are given as [6–8]:

$$w(x,t)=\sum_{m=1}^{\infty}W_m\sin\left(\frac{n\pi}{L}x\right)$$

(6.19)

$$\phi(x,t)=\sum_{m=1}^{\infty}\Phi_m\cos\left(\frac{n\pi}{L}x\right)$$

(6.20)

In which, W_m and Φ_m are the amplitudes of transverse displacement and rotation of cross-section, respectively.

Substituting Eq. (6.19) and Eq. (6.20) into Eq. (6.17) and Eq. (6.18) yields:

$$\begin{bmatrix} -k_s\dfrac{\pi(\tilde{d})^2}{4}\dfrac{\tilde{E}}{2(1+\tilde{v})}\left(\dfrac{n\pi}{\tilde{L}}\right)^2+P\left(\dfrac{n\pi}{\tilde{L}}\right)^2 & -k_s\dfrac{\pi(\tilde{d})^2}{4}\dfrac{\tilde{E}}{2(1+\tilde{v})}\left(\dfrac{n\pi}{\tilde{L}}\right) \\ -k_s\dfrac{\pi(\tilde{d})^2}{4}\dfrac{\tilde{E}}{2(1+\tilde{v})}\left(\dfrac{n\pi}{\tilde{L}}\right)-\mu P\left(\dfrac{n\pi}{\tilde{L}}\right)^3 & -\tilde{E}\dfrac{\pi(\tilde{d})^4}{64}\left(\dfrac{n\pi}{\tilde{L}}\right)^2-k_s\dfrac{\pi(\tilde{d})^2}{4}\dfrac{\tilde{E}}{2(1+\tilde{v})} \end{bmatrix}$$

$$\begin{bmatrix} W_m \\ \Phi_m \end{bmatrix}=0$$

(6.21)

Where,

$$\tilde{d}(\alpha,\beta)=gfn(\bar{d},\sigma_1,\sigma_1)=\left(\left(2\beta\sqrt{-2\sigma_1^2\ln(\alpha)}\right)+\bar{d}-\sqrt{-2\sigma_1^2\ln(\alpha)}\right),$$

$$\tilde{L}(\alpha, \beta) = gfn(\bar{L}, \sigma_2, \sigma_2) = \left[\left(2\beta\sqrt{-2\sigma_2^2 \ln(\alpha)}\right) + \bar{L} - \sqrt{-2\sigma_2^2 \ln(\alpha)}\right],$$

$$\tilde{E}(\alpha, \beta) = gfn(\bar{E}, \sigma_3, \sigma_3) = \left[\left(2\beta\sqrt{-2\sigma_3^2 \ln(\alpha)}\right) + \bar{E} - \sqrt{-2\sigma_3^2 \ln(\alpha)}\right], \text{ with } \alpha \in (0 \; 1]$$

and $\beta \in [0 \; 1]$.

These are nothing but the double parametric forms of the Symmetric Gaussian Fuzzy Number.

Taking determinant of the matrix obtained in Eq. (6.21), the buckling loads of the nanobeam can be found out for uncertain parameters. It is worth mentioning that the present uncertain model can be debased into the deterministic model by substituting $\alpha = 1$ in Eq. (6.21).

6.4 NUMERICAL RESULTS AND DISCUSSION

The critical buckling load (P_1) or (P_{cr}) and other higher buckling loads (P_n), $n = 2, 3, 4,$ etc., are calculated from Eq. (6.21) with uncertain parameters. The modal values with their standard deviations, i.e., Gaussian fuzzy numbers associated with uncertain parameters, are presented in Table 6.1, and graphical presentation of uncertain parameters is also illustrated in Figure 6.1. It may be noted that the standard deviation corresponds to a worst case of $\pm 15\%$ for the diameter, whereas $\pm 9\%$ and $\pm 6\%$ are associated with the length and Young's modulus, respectively. For different values of α, different intervals are generated, and for each interval, lower bound (LB) and upper bound (UB) of buckling loads are determined by incorporating the double parametric form in Navier's approach. The intervals that are considered in this chapter for the investigation are demonstrated in Table 6.2. For computational purpose, following values of the parameters are adapted from Wang et al. [3] as:

$$E = 1TPa, \; G = \frac{E}{2(1+v)}, \; v = 0.19, \; d = 1, \text{ and } I = \frac{\pi d^4}{64}.$$

Also, the study is conducted in four ways, i.e., the results are generated by considering (i) the diameter (d) as fuzzy or uncertain, (ii) the length (L) as fuzzy, (iii) Young's modulus (E) as fuzzy, and, finally, (iv) d, L, and, E are as uncertain parameters.

Table 6.1
Assigning of Symmetric Gaussian Fuzzy Number (SGFN) to Uncertain Parameters

Parameters	Modal Value	Sym. Gaussian Fuzzy Number
Diameter (d)	$\bar{d} = 1\,nm$	$\tilde{d} = sgfn(\bar{d}, 5\%\bar{d}, 5\%\bar{d})$
Length (L)	$\bar{L} = 10\,nm$	$\tilde{L} = sgfn(\bar{L}, 3\%\bar{L}, 3\%\bar{L})$
Young's Modulus (E)	$\bar{E} = 1TPa$	$\tilde{E} = sgfn(\bar{E}, 2\%\bar{E}, 2\%\bar{E})$

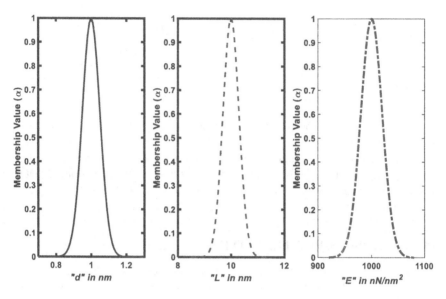

FIGURE 6.1 Symmetric Gaussian Fuzzy Numbers (SGFNs) associated with "diameter (*d*)", "length (*L*)", and "Young's modulus (*E*)".

Table 6.2

Intervals for Different α, for Uncertain Parameters

α	$\tilde{L} = \left[\underline{L}(\alpha), \overline{L}(\alpha)\right]$	$\tilde{d} = \left[\underline{d}(\alpha), \overline{d}(\alpha)\right]$	$\tilde{E} = \left[\underline{E}(\alpha), \overline{E}(\alpha)\right]$
0.1	[9.3562 10.6438]	[0.8927 1.1073]	[957.0807 1042.9]
0.2	[9.4618 10.5382]	[0.9103 1.0897]	[964.1175 1035.9]
0.3	[9.5345 10.4655]	[0.9224 1.0776]	[968.9649 1031]
0.4	[9.5939 10.4061]	[0.9323 1.0677]	[972.9254 1027.1]
0.5	[9.6468 10.3532]	[0.9411 1.0589]	[976.4518 1023.5]
0.6	[9.6968 10.3032]	[0.9495 1.0505]	[979.7846 1020.2]
0.7	[9.7466 10.2534]	[0.9578 1.0422]	[983.1080 1016.9]
0.8	[9.7996 10.2004]	[0.9666 1.0334]	[986.6391 1013.4]
0.9	[9.8623 10.1377]	[0.9770 1.0230]	[990.8191 1009.2]
1	10 (Modal value of *L*)	1 (Modal value of *d*)	1000 (Modal value of *E*)

6.4.1 VALIDATION OF THE PROPOSED MODEL

The proposed model is validated by comparing the results of the hinged–hinged (H-H) boundary condition with the deterministic model. For this purpose, the membership value α has been taken as 1, and the result obtained from the present model is compared with Wang et al. [3]. All the parameters are kept as same as Wang et al. [3], and the validation is shown in Table 6.3. From these results, it is quite evident that the present uncertain model exhibits an excellent agreement with Wang et al. [3], in specific cases.

Table 6.3

Validation of the Uncertain Model for $\alpha = 1$ with the Deterministic Model of Wang et al. [3]

$(e_0 a)$ in nm	Present Model by Setting $\alpha = 1$				Wang et al. [3]			
	$L = 10$	$L = 12$	$L = 14$	$L = 16$	$L = 10$	$L = 12$	$L = 14$	$L = 16$
0	4.7670	3.3267	2.4514	1.8805	4.7670	3.3267	2.4514	1.8805
0.5	4.654	3.2713	2.4212	1.8626	4.654	3.2713	2.4212	1.8626
1	4.3450	3.1156	2.3348	1.8111	4.3450	3.1156	2.3348	1.8111
1.5	3.9121	2.8865	2.2038	1.7313	3.9121	2.8865	2.2038	1.7313
2	3.4333	2.6172	2.0432	1.6306	3.4333	2.6172	2.0432	1.6306

6.4.2 DOUBLE PARAMETRIC SOLUTION VERSUS MONTE CARLO SIMULATION TECHNIQUE

This subsection is devoted to the comparison of non-probabilistic method, i.e., double parametric form of solution, with the Monte Carlo Simulation Technique. In the double parametric form of solution, the uncertain parameters are taken as Symmetric Gaussian Fuzzy Numbers (SGFNs), as given in Table 6.1. Further, by applying α cut technique or single parametric form, the uncertain parameters degenerate into interval parameters as listed in Table 6.2. This interval forms of uncertain parameters are later converted into crisp form or deterministic form by incorporating double parametric form as explained in the preliminaries. In case of Monte Carlo Simulation, the Symmetric Gaussian Fuzzy Numbers are converted into interval form by using α cut technique, then, 1,000 random points are generated for interval form of uncertain parameters from their lower bounds to upper bounds. Then, these random values of uncertain parameters are used in deterministic analysis of critical buckling load. By repeating deterministic analysis for all the random points, lower bound and upper bound of critical buckling loads are computed. For computational purpose, we have considered three different cases such as (i) the diameter (d) as fuzzy or uncertain, (ii) the length (L) as fuzzy, and (iii) Young's modulus (E) as fuzzy. Table 6.4, Table 6.5, and Table 6.6 respectively represent the tabular solutions for diameter, length, and Young's modulus as uncertain parameters. Likewise, Figures 6.2–6.4 represent the graphical results in the order of tabular results. From these results, it is quite evident that the results obtained by double parametric form and Monte Carlo Simulation Technique are showing an excellent agreement in three cases. Another interesting observation is that the lower bound (LB) of critical buckling load calculated from Monte Carlo technique is little bit higher than that of the lower bound (LB) of critical buckling load obtained from double parametric form. However, this trend is reversed in case of upper bound (UB).

Table 6.4

Comparison of the Solution Obtained by Double Parametric Form versus Monte Carlo Simulation Technique for Critical Buckling Load with $\tilde{d} = gfn\left(\bar{d}, 5\%\,\bar{d}, 5\%\,\bar{d}\right)$ and $e_0 a = 1$

α	$\tilde{d}$		P_{cr} in nN (Proposed Method)[+]		P_{cr} in nN (MCS)[#]		Error (%)	
	LB	UB	LB	UB	LB	UB	LB	UB
0.1	0.8927	1.1073	2.7676	6.5105	2.7688	6.5025	0.0276	0.1841
0.2	0.9103	1.0897	2.9909	6.1100	2.9916	6.1066	0.0161	0.0783
0.3	0.9224	1.0776	3.1524	5.8449	3.1529	5.8441	0.0115	0.0184
0.4	0.9323	1.0677	3.2891	5.6348	3.2938	5.6347	0.1082	0.0023
0.5	0.9411	1.0589	3.4144	5.4525	3.4170	5.4504	0.0598	0.0483
0.6	0.9495	1.0505	3.5361	5.2842	3.5371	5.2737	0.0230	0.2417
0.7	0.9578	1.0422	3.6607	5.1203	3.6619	5.1193	0.0276	0.0230
0.8	0.9666	1.0334	3.7966	4.9504	3.7967	4.9501	0.0023	0.0069
0.9	0.9770	1.0230	3.9623	4.7546	3.9624	4.7528	0.0023	0.0414
1	1	1	4.3450	4.3450	4.3450	4.3450	0	0

[#] Time taken for LB and UB by MCS: 49.188114 seconds
[+] Time taken for LB and UB by the proposed method: 7.097949 seconds

Table 6.5

Comparison of the Solution Obtained by Double Parametric Form versus Monte Carlo Simulation Technique for Critical Buckling Load with $\tilde{L} = gfn\left(\bar{L}, 3\%\,\bar{L}, 3\%\,\bar{L}\right)$ and $e_0 a = 1$

α	$\tilde{L}$		P_{cr} in nN (Proposed Method)[+]		P_{cr} in nN (MCS)[#]		Error (%)	
	LB	UB	LB	UB	LB	UB	LB	UB
0.1	9.3562	10.6438	3.8823	4.8917	3.8835	4.8908	0.0276	0.0207
0.2	9.4618	10.5382	3.9531	4.7955	3.9533	4.7953	0.0046	0.0046
0.3	9.5345	10.4655	4.0030	4.7309	4.0068	4.7308	0.0875	0.0023
0.4	9.5939	10.4061	4.0444	4.6789	4.0450	4.6789	0.0138	0
0.5	9.6468	10.3532	4.0818	4.6334	4.0821	4.6334	0.0069	0
0.6	9.6968	10.3032	4.1177	4.5909	4.1184	4.5902	0.0161	0.0161
0.7	9.7466	10.2534	4.1538	4.5492	4.1542	4.5491	0.0092	0.0023
0.8	9.7996	10.2004	4.1928	4.5054	4.1932	4.5048	0.0092	0.0138
0.9	9.8623	10.1377	4.2396	4.4543	4.2398	4.4542	0.0046	0.0023
1	10	10	4.3450	4.3450	4.3450	4.3450	0	0

[#] Time taken for LB and UB by MCS: 55.188114 seconds
[+] Time taken for LB and UB by the proposed method: 6.967949 seconds

Table 6.6

Comparison of the Solution Obtained by Double Parametric Form versus Monte Carlo Simulation Technique for Critical Buckling Load with $\tilde{E} = gfn\left(\bar{d}, 2\%\,\bar{E}, 2\%\,\bar{E}\right)$ and $e_0a = 1$

	$\tilde{E}$		P_{cr} in nN (Proposed Method)[+]		P_{cr} in nN (MCS)[#]		Error (%)	
α	LB	UB	LB	UB	LB	UB	LB	UB
0.1	957.0807	1042.9	4.1585	4.5315	4.1587	4.5313	0.0046	0.0046
0.2	964.1175	1035.9	4.1891	4.5009	4.1892	4.5008	0.0023	0.0023
0.3	968.9649	1031	4.2102	4.4799	4.2103	4.4798	0.0023	0.0023
0.4	972.9254	1027.1	4.2274	4.4627	4.2277	4.4626	0.0069	0.0023
0.5	976.4518	1023.5	4.2427	4.4473	4.2427	4.4470	0	0.0069
0.6	979.7846	1020.2	4.2572	4.4329	4.2574	4.4325	0.0046	0.0092
0.7	983.1080	1016.9	4.2716	4.4184	4.2718	4.4182	0.0046	0.0046
0.8	986.6391	1013.4	4.2870	4.4031	4.2870	4.4030	0	0.0023
0.9	990.8191	1009.2	4.3051	4.3849	4.3051	4.3844	0	0.0115
1	1000	1000	4.3450	4.3450	4.3450	4.3450	0	0

[#] Time taken for LB and UB by MCS: 65.301382 seconds
[+] Time taken for LB and UB by the proposed method: 7.053047seconds

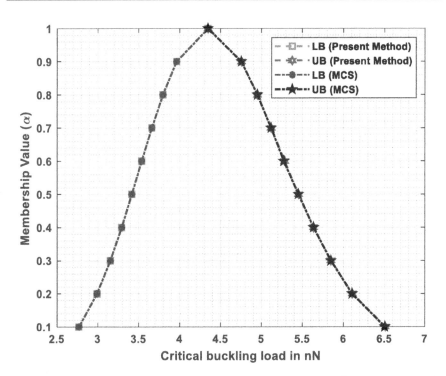

FIGURE 6.2 Present method versus MCS for critical buckling load with $\tilde{d} = sgfn\left(1, 0.05, 0.05\right)$.

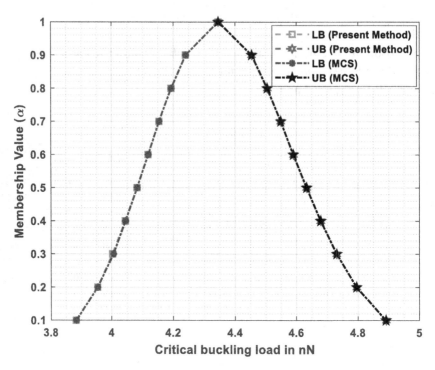

FIGURE 6.3   Present method versus MCS for critical buckling load with $\tilde{L} = sgfn(10, 0.3, 0.3)$.

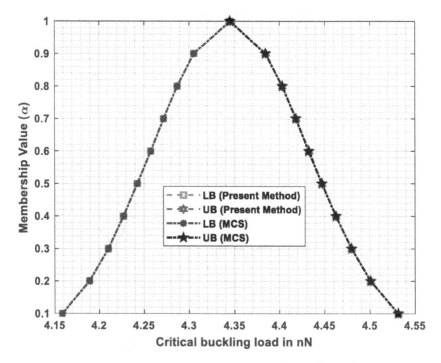

FIGURE 6.4   Present method versus MCS for critical buckling load with $\tilde{E} = sgfn(1, 0.02, 0.02)$.

6.4.3 Effects of Material Uncertainties

Material uncertainties in nanostructures are induced due to the defects in atomic structure and manufacturing abnormality, which affects the dynamical characteristics considerably. In this section, we will analyze the effects of material uncertainties on the stability of nanobeam. The uncertain input variables are considered as Symmetric Gaussian Fuzzy Number (SGFN) as given in Table 6.1 and Figure 6.1. The double parametric form is incorporated with Navier's technique to calculate the buckling loads for Timoshenko nanobeam, which are demonstrated in Tables 6.7–6.10 as tabular results for all the four cases: (i) The diameter (d) as fuzzy or uncertain, (ii) the length (L) as fuzzy, (iii) Young's modulus (E) as fuzzy, and, finally, (iv) d, L, and E are as uncertain parameters. The graphical results for critical buckling loads are illustrated in Figures 6.5–6.8. From the graphical results, it can be perceived that the output for critical buckling loads is being obtained as Gaussian fuzzy numbers. Further, a comparative study has been conducted by considering the first four buckling loads as the Gaussian fuzzy number, and the same has been presented as graphical results, which are depicted in Figures 6.9–6.12. The spreads or fuzziness of Gaussian fuzzy output of buckling loads is the highest in case of uncertain diameter, whereas the fuzziness of buckling loads is the lowest in case of uncertain Young's modulus as input. Also, in case of higher buckling loads, the fuzziness values of "all uncertain parameters" as input and "uncertain diameter" as input are matching with each other, and the same pattern is observed in the case of uncertain Young's modulus as input and "uncertain length" as input.

Table 6.7

Lower Bound and Upper Bound of Buckling Loads in nN for Different Values of α with $\bar{d} = sgfn(1, 0.05, 0.05)$ and $e_0 a = 1 nm$

α	$\beta = 0$ (Lower Bound)				$\beta = 1$ (Upper Bound)			
	$\underline{P}_{cr}$	$\underline{P}_2$	$\underline{P}_3$	$\underline{P}_4$	$\bar{P}_{cr}$	$\bar{P}_2$	$\bar{P}_3$	$\bar{P}_4$
0.1	2.7676	8.5065	13.8091	17.6627	6.5105	19.7542	31.6930	40.1956
0.2	2.9909	9.1840	14.8957	19.0402	6.1100	18.5601	29.8082	37.8330
0.3	3.1524	9.6732	15.6794	20.0330	5.8449	17.7689	28.5576	36.2640
0.4	3.2891	10.0870	16.3417	20.8713	5.6348	17.1409	27.5641	35.0168
0.5	3.4144	10.4662	16.9482	21.6388	5.4525	16.5956	26.7007	33.9322
0.6	3.5361	10.8341	17.5363	22.3826	5.2842	16.0919	25.9027	32.9293
0.7	3.6607	11.2104	18.1374	23.1424	5.1203	15.6009	25.1242	31.9504
0.8	3.7966	11.6205	18.7922	23.9697	4.9504	15.0914	24.3157	30.9333
0.9	3.9623	12.1203	19.5893	24.9764	4.7546	14.5040	23.3830	29.7593
1	4.3450	13.2729	21.4255	27.2930	4.3450	13.2729	21.4255	27.2930

Table 6.8

Lower Bound and Upper Bound of Buckling Loads in nN for Different Values of α with $\tilde{L} = sgfn(10, 0.3, 0.3)$ and $e_0 a = 1\,nm$

α	$\beta = 0$ (Lower Bound)				$\beta = 1$ (Upper Bound)			
	$\underline{P}_{cr}$	$\underline{P}_2$	$\underline{P}_3$	$\underline{P}_4$	$\overline{P}_{cr}$	$\overline{P}_2$	$\overline{P}_3$	$\overline{P}_4$
0.1	3.8823	12.1655	20.1121	26.0729	4.8917	14.5115	22.8231	28.5454
0.2	3.9531	12.3387	20.3217	26.2705	4.7955	14.2988	22.5882	28.3382
0.3	4.0030	12.4599	20.4674	26.4072	4.7309	14.1545	22.4277	28.1958
0.4	4.0444	12.5600	20.5872	26.5192	4.6789	14.0380	22.2974	28.0797
0.5	4.0818	12.6500	20.6945	26.6192	4.6334	13.9353	22.1820	27.9766
0.6	4.1177	12.7359	20.7964	26.7139	4.5909	13.8391	22.0734	27.8793
0.7	4.1538	12.8222	20.8985	26.8086	4.5492	13.7439	21.9656	27.7824
0.8	4.1928	12.9148	21.0076	26.9094	4.5054	13.6437	21.8516	27.6797
0.9	4.2396	13.0255	21.1375	27.0291	4.4543	13.5263	21.7174	27.5584
1	4.3450	13.2729	21.4255	27.2930	4.3450	13.2729	21.4255	27.2930

Table 6.9

Lower Bound and Upper Bound of Buckling Loads in nN for Different Values of α with $\tilde{E} = sgfn(1, 0.02, 0.02)$ and $e_0 a = 1\,nm$

α	$\beta = 0$ (Lower Bound)				$\beta = 1$ (Upper Bound)			
	$\underline{P}_{cr}$	$\underline{P}_2$	$\underline{P}_3$	$\underline{P}_4$	$\overline{P}_{cr}$	$\overline{P}_2$	$\overline{P}_3$	$\overline{P}_4$
0.1	4.1585	12.7033	20.5060	26.1216	4.5315	13.8426	22.3451	28.4644
0.2	4.1891	12.7967	20.6567	26.3136	4.5009	13.7492	22.1943	28.2723
0.3	4.2102	12.8610	20.7606	26.4459	4.4799	13.6849	22.0905	28.1400
0.4	4.2274	12.9136	20.8455	26.5540	4.4627	13.6323	22.0056	28.0319
0.5	4.2427	12.9604	20.9210	26.6503	4.4473	13.5855	21.9301	27.9357
0.6	4.2572	13.0046	20.9924	26.7412	4.4329	13.5412	21.8587	27.8447
0.7	4.2716	13.0487	21.0636	26.8319	4.4184	13.4971	21.7875	27.7540
0.8	4.2870	13.0956	21.1393	26.9283	4.4031	13.4503	21.7118	27.6576
0.9	4.3051	13.1511	21.2288	27.0424	4.3849	13.3948	21.6222	27.5436
1	4.3450	13.2729	21.4255	27.2930	4.3450	13.2729	21.4255	27.2930

Table 6.10

Lower Bound and Upper Bound of Buckling loads in nN for Different Values of α with $\tilde{L} = sgfn(10, 0.3, 0.3)$, $\tilde{d} = sgfn(1, 0.05, 0.05)$, $\tilde{E} = sgfn(1, 0.02, 0.02)$, and $e_0 a = 1nm$

α	$\beta = 0$ (Lower Bound)				$\beta = 1$ (Upper Bound)			
	$\underline{P}_{cr}$	$\underline{P}_2$	$\underline{P}_3$	$\underline{P}_4$	$\overline{P}_{cr}$	$\overline{P}_2$	$\overline{P}_3$	$\overline{P}_4$
0.1	2.9833	8.9087	14.0922	17.6959	6.0690	18.8989	31.0576	40.0832
0.2	3.1834	9.5445	15.1508	19.0716	5.7597	17.8834	29.3071	37.7463
0.3	3.3265	10.0001	15.9116	20.0625	5.5530	17.2056	28.1416	36.1931
0.4	3.4464	10.3831	16.5526	20.8989	5.3877	16.6650	27.2133	34.9577
0.5	3.5557	10.7325	17.1384	21.6642	5.2434	16.1933	26.4047	33.8830
0.6	3.6610	11.0699	17.7052	22.4056	5.1094	15.7559	25.6559	32.8888
0.7	3.7681	11.4136	18.2832	23.1626	4.9780	15.3278	24.9240	31.9180
0.8	3.8842	11.7866	18.9116	23.9867	4.8409	14.8816	24.1623	30.9088
0.9	4.0247	12.2387	19.6748	24.9887	4.6819	14.3649	23.2815	29.7434
1	4.3450	13.2729	21.4255	27.2930	4.3450	13.2729	21.4255	27.2930

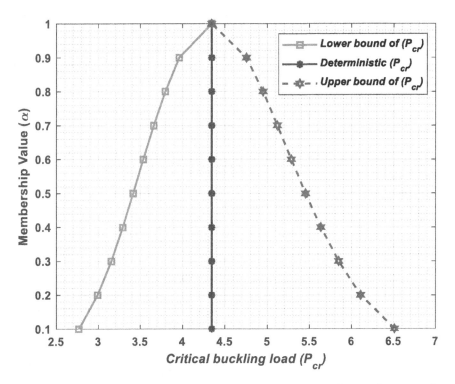

FIGURE 6.5 Gaussian fuzzy output for critical buckling load with $\tilde{d} = sgfn(1, 0.05, 0.05)$.

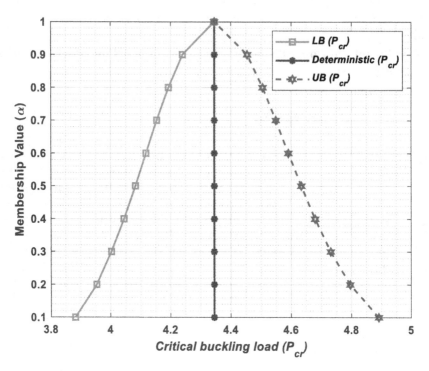

FIGURE 6.6 Gaussian fuzzy output for critical buckling load with $\tilde{L} = sgfn(10, 0.3, 0.3)$.

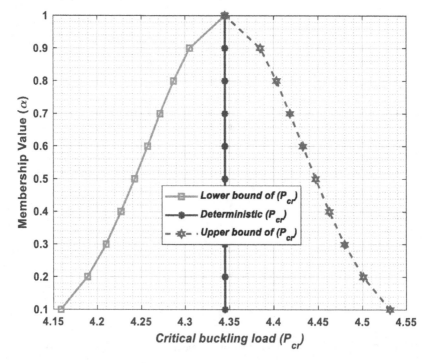

FIGURE 6.7 Gaussian fuzzy output for critical buckling load with $\tilde{E} = sgfn(1, 0.02, 0.02)$.

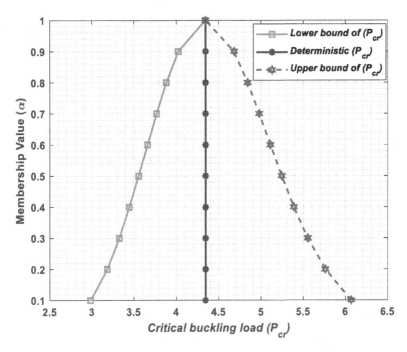

FIGURE 6.8 Gaussian fuzzy output for critical buckling load with $\tilde{L} = sgfn(10, 0.3, 0.3)$, $\tilde{d} = sgfn(1, 0.05, 0.05)$, and $\tilde{E} = sgfn(1, 0.02, 0.02)$.

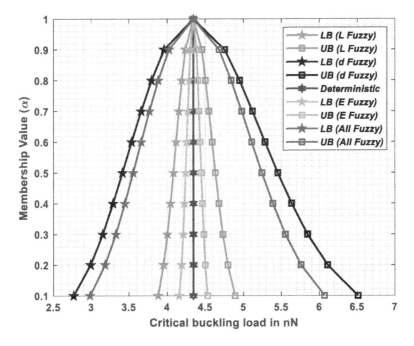

FIGURE 6.9 Comparisons of Gaussian fuzzy output for critical buckling load.

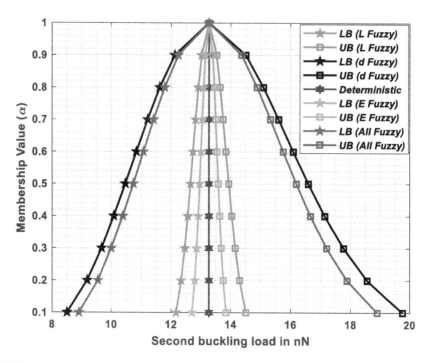

FIGURE 6.10   Comparisons of Gaussian fuzzy output for second buckling load.

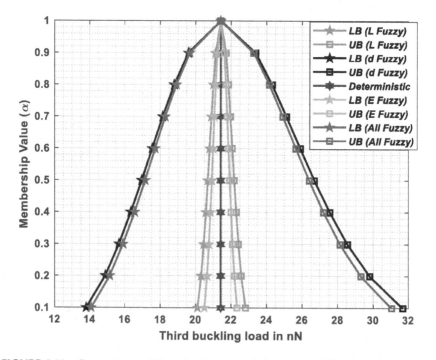

FIGURE 6.11   Comparisons of Gaussian fuzzy output for third buckling load.

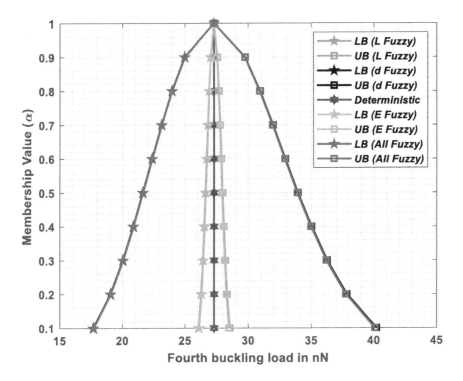

FIGURE 6.12   Comparisons of Gaussian fuzzy output for fourth buckling load.

6.5   CONCLUDING REMARKS

In this chapter, a non-probabilistic method, specifically the double parametric form-based Navier's approach, has been employed to investigate the effects of material uncertainties on the stability analysis of Timoshenko nanobeam. The uncertainties in material properties are assumed to be represented by a unique fuzzy number, namely Symmetric Gaussian Fuzzy Number (SGFN), and are associated with the diameter, length, and Young's modulus of the nanobeam. The double parametric form-based Navier's method has been employed to generate the lower bound (LB) and upper bound (UB) of the buckling loads of the uncertain models. The results obtained by the uncertain model have been verified with the deterministic model in special cases and the Monte Carlo Simulation Technique, revealing a strong agreement. The main observations are summarized as follows:

- The Lower Bound (LB) of the critical buckling load calculated from the Monte Carlo technique is found to be slightly higher than that of the LB of the critical buckling load obtained from the double parametric form-based solution. However, this trend is reversed in the case of the upper bound (UB).
- The output of the buckling loads has been obtained as Gaussian fuzzy numbers whereas the input was given in terms of Symmetric Gaussian Fuzzy Numbers.

- The spreads or fuzziness of Gaussian fuzzy output of buckling loads is the highest in the case of uncertain diameter whereas the fuzziness of buckling loads is the lowest in the case of uncertain Young's modulus as input.
- In the case of higher buckling loads, the fuzziness of "all uncertain parameters" as input and "uncertain diameter" as input matches with each other, and the same pattern is observed in the case of uncertain Young's modulus as input and "uncertain length" as input.

BIBILIOGRAPHY

[1] Reddy JN (2007) Nonlocal theories for bending, buckling and vibration of beams. International Journal of Engineering Science 45(2–8):288–307

[2] Wang CM, Reddy JN, Lee KH (2000) Shear deformable beams and plates: Relationships with classical solutions. Elsevier

[3] Wang CM, Zhang YY, Ramesh SS, Kitipornchai S (2006) Buckling analysis of micro- and nano-rods/tubes based on nonlocal Timoshenko beam theory. Journal of Physics D: Applied Physics 39(17):3904

[4] Malikan M, Eremeyev VA. (2020 Apr) On the dynamics of a Visco—Piezo—Flexoelectric nanobeam. Symmetry 12(4):643

[5] Eringen AC (1972) Nonlocal polar elastic continua. International Journal of Engineering Science 10:1–16

[6] Malikan M, Dastjerdi S (2018) Analytical buckling of FG nanobeams on the basis of a new one variable first-order shear deformation beam theory. International Journal of Engineering Applied Science 10:21–34

[7] Malikan M, Dimitri R, Tornabene F (2019) Transient response of oscillated carbon nanotubes with an internal and external damping. Composites Part B Engineering 158:198–205

[8] Malikan M, Nguyen VB, Tornabene F (2018) Damped forced vibration analysis of single-walled carbon nanotubes resting on viscoelastic foundation in thermal environment using nonlocal strain gradient theory. Engineering Science and Technology an International Journal 21:778–786

7 Vibrations of Functionally Graded Beam with Uncertainty

7.1 INTRODUCTION

This chapter delves into a comprehensive exploration of the profound impact that material uncertainties exert on the vibration characteristics inherent to functionally graded (FG) beams. Within this study, these uncertainties are intricately linked to the Young's modulus and material density of the metal constituent, represented by specialized fuzzy numbers, namely Triangular Fuzzy Numbers (TFNs). Through a meticulous integration of the Euler–Bernoulli beam theory alongside a double parametric form of fuzzy numbers, the governing equations governing the vibrational behavior of the uncertain model are meticulously derived. Subsequently, employing the Rayleigh–Ritz method, the outcomes for both the lower bound (LB) and upper bound (UB) of the natural frequencies concerning the uncertain models under simply supported–simply supported (SS), clamped–simply supported (CS), and clamped–clamped (CC) boundary conditions are meticulously computed. To ensure the robustness and reliability of the uncertain model, a thorough validation against the deterministic model is rigorously conducted, demonstrating a commendable level of agreement. Furthermore, a comprehensive parametric exploration is undertaken to probe into the intricacies of fuzziness or spreads pertaining to the natural frequency in relation to a diverse array of uncertain parameters, thereby offering invaluable insights into the multifaceted nature of uncertainty within the vibrational domain.

7.2 MATHEMATICAL FORMULATION OF THE PROPOSED PROBLEM

According to the rule of the mixture, the material property P of the FG nanobeam can be written as [1–5]:

$$P = P_c V_c + P_m V_m \tag{7.1}$$

Here, P_c and V_c represent the material property and volume fraction, respectively, for the ceramic constituent, while P_m and V_m denote the material property and volume fraction of the metal constituent. According to the power-law variation model, the volume fractions of the ceramic and metal components are formulated as described in References [1–5]:

$$V_c = \left(\frac{z}{h} + \frac{1}{2}\right)^k \tag{7.2}$$

DOI: 10.1201/9781003303107-7

$$V_m = 1 - \left(\frac{z}{h} + \frac{1}{2}\right)^k \qquad (7.3)$$

Where k represents a non-negative parameter known as the power-law exponent, which governs the distribution of material throughout the thickness of the nanobeam, while z denotes the distance from the mid-plane of the FG nanobeam. Utilizing Equations (7.1), (7.2), and (7.3), the material properties of the FG nanobeam can be expressed, as referenced in [1–5].

$$P = \left(P_c - P_m\right)\left(\frac{z}{h} + \frac{1}{2}\right)^k + P_m \qquad (7.4)$$

The Young's modulus $E(z)$ and material density $\rho(z)$ of the FG beam can be presented mathematically as:

$$E(z) = \left(E_c - E_m\right)\left(\frac{z}{h} + \frac{1}{2}\right)^k + E_m \qquad (7.5)$$

$$\rho(z) = \left(\rho_c - \rho_m\right)\left(\frac{z}{h} + \frac{1}{2}\right)^k + \rho_m \qquad (7.6)$$

Given that uncertainties are attributed to both the Young's modulus and material density of the metal constituent, we integrate the double parametric form, as outlined in the preliminaries. Consequently, the Triangular Fuzzy Numbers $\tilde{E}_m = \left(e_1, e_2, e_3\right)$ and $\tilde{\rho}_m = \left(\rho_1, \rho_2, \rho_3\right)$ can be represented in double parametric form as:

$$\tilde{E}_m\left(\alpha_1, \beta_1\right) = \left(e_1 - e_3\right)\alpha_1\beta_1 + \left(e_3 - e_1\right)\beta_1 + \left(e_2 - e_1\right)\alpha_1 + e_1, \quad \alpha_1, \beta_1 \in \left[0 \ 1\right] \quad (7.7)$$

$$\tilde{\rho}_m\left(\alpha_2, \beta_2\right) = \left(\rho_1 - \rho_3\right)\alpha_2\beta_2 + \left(\rho_3 - \rho_1\right)\beta_2 + \left(\rho_2 - \rho_1\right)\alpha_2 + \rho_1, \quad \alpha_2, \beta_2 \in \left[0 \ 1\right] (7.8)$$

Using Eq. (7.7) and Eq. (7.8) in Eq. (7.5) and Eq. (7.6), the Young's modulus and material density of the FG beam can be expressed as:

$$E\left(z, \alpha_1, \beta_1\right) = \left(E_c - \tilde{E}_m\left(\alpha_1, \beta_1\right)\right)\left(\frac{z}{h} + \frac{1}{2}\right)^k + \tilde{E}_m\left(\alpha_1, \beta_1\right) \qquad (7.9)$$

$$\rho\left(z, \alpha_2, \beta_2\right) = \left(\rho_c - \tilde{\rho}_m\left(\alpha_2, \beta_2\right)\right)\left(\frac{z}{h} + \frac{1}{2}\right)^k + \tilde{\rho}_m\left(\alpha_2, \beta_2\right) \qquad (7.10)$$

In accordance with the Euler–Bernoulli beam theory, the displacement field can be expressed as [6]:

$$u_1\left(x, z, t\right) = u\left(x, t\right) - z\left(\frac{\partial w}{\partial x}\right)$$

$$u_2\left(x, z, t\right) = 0 \qquad (7.11)$$

$$u_3\left(x, z, t\right) = w\left(x, t\right)$$

Where $u(x,t)$ and $w(x,t)$ represent the axial and transverse displacements on the mid-plane of the FG beam, respectively. The strain–displacement relation of the FG beam is stated as:

$$\varepsilon_{xx} = \frac{\partial u_1(x,z,t)}{\partial x} = \frac{\partial u(x,t)}{\partial x} - z\frac{\partial^2 w(x,t)}{\partial x^2} \tag{7.12}$$

The stress component of the FG beam as generalized Hooke's law may be given as [1–2]:

$$\sigma_{xx} = Q_{11}\varepsilon_{xx} = \left(\frac{E(z,\alpha_1,\beta_1)}{1-v^2}\right)\varepsilon_{xx} \tag{7.13}$$

The strain energy (S) of the FG beam is stated as [1–2, 7]:

$$
\begin{aligned}
S &= \frac{1}{2}\int_0^L \int_A (\sigma_{xx}\varepsilon_{xx}) \, dA \, dx \\
&= \frac{1}{2}\int_0^L \int_A \sigma_{xx}\left[\frac{\partial u(x,t)}{\partial x} - z\frac{\partial^2 w(x,t)}{\partial x^2}\right] dA \, dx \\
&= \frac{1}{2}\int_0^L \left[N\left(\frac{\partial u(x,t)}{\partial x}\right) - M\left(\frac{\partial^2 w(x,t)}{\partial x^2}\right)\right] dx
\end{aligned} \tag{7.14}
$$

Where the stress resultants $(N,M) = \int_A (\sigma_{xx}, z\sigma_{xx}) \, dA$.

The kinetic energy (T) of the FG beam can be stated as [1–2, 7]:

$$
\begin{aligned}
T &= \frac{1}{2}\int_0^L \int_A \rho(z,\alpha_2,\beta_2)\left[\left(\frac{\partial u_1}{\partial t}\right)^2 + \left(\frac{\partial u_2}{\partial t}\right)^2 + \left(\frac{\partial u_3}{\partial t}\right)^2\right] dA \, dx \\
&= \frac{1}{2}\int_0^L \int_A \rho(z,\alpha_2,\beta_2)\left[\left(\frac{\partial u}{\partial t} - z\frac{\partial^2 w}{\partial x \partial t}\right)^2 + \left(\frac{\partial w}{\partial t}\right)^2\right] dA \, dx \\
&= \frac{1}{2}\int_0^L \left[I_0\left(\left(\frac{\partial u}{\partial t}\right)^2 + \left(\frac{\partial w}{\partial t}\right)^2\right) - 2I_1\left(\frac{\partial u}{\partial t}\right)\left(\frac{\partial^2 w}{\partial x \partial t}\right) + I_2\left(\frac{\partial^2 w}{\partial x \partial t}\right)^2\right] dx
\end{aligned} \tag{7.15}
$$

In which $(I_0, I_1, I_2) = \int_A \rho(z,\alpha_2,\beta_2)(1, z, z^2) \, dA$ are the mass moment of inertias.

Multiplying Eq. (7.13) by dA and $z\,dA$ and integrating over the area of cross-section of the FG beam, the local stress resultants can be written as [7]:

$$N = A_{11}\left(\frac{\partial u}{\partial x}\right) - B_{11}\left(\frac{\partial^2 w}{\partial x^2}\right) \tag{7.16}$$

$$M = B_{11}\left(\frac{\partial u}{\partial x}\right) - D_{11}\left(\frac{\partial^2 w}{\partial x^2}\right), \tag{7.17}$$

where $\left(A_{11}, B_{11}, D_{11}\right) = \int_A Q_{11}\left(1, z, z^2\right)dA = \int_A \left(\frac{E(z,\alpha_1,\beta_1)}{1-v^2}\right)\left(1, z, z^2\right)dA$ are the stiff-nesses of the FG beam. Substituting Eq. (7.16) and Eq. (7.17) in Eq. (7.14), the strain energy and kinetic energy for the FG beam can be depicted as [7]:

$$S = \frac{1}{2}\int_0^L\left[A_{11}\left(\frac{\partial u}{\partial x}\right)^2 - 2B_{11}\left(\frac{\partial u}{\partial x}\right)\left(\frac{\partial^2 w}{\partial x^2}\right) + D_{11}\left(\frac{\partial^2 w}{\partial x^2}\right)^2\right]dx \tag{7.18}$$

$$T = \frac{1}{2}\int_0^L\left[I_0\left(\left(\frac{\partial u}{\partial t}\right)^2 + \left(\frac{\partial w}{\partial t}\right)^2\right) - 2I_1\left(\frac{\partial u}{\partial t}\right)\left(\frac{\partial^2 w}{\partial x \partial t}\right) + I_2\left(\frac{\partial^2 w}{\partial x \partial t}\right)^2\right]dx \tag{7.19}$$

Assuming the motion of the FG beam as sinusoidal, i.e., plugging $u(x,t) = U(x)\cos(\omega t)$ and $w(x,t) = W(x)\cos(\omega t)$, the maximum strain energy (S_{max}) and kinetic energy (T_{max}) for the FG beam can be obtained as [1–2, 8]:

$$S_{max} = \frac{1}{2}\int_0^L\left[A_{11}\left(\frac{dU}{dx}\right)^2 - 2B_{11}\left(\frac{dU}{dx}\right)\left(\frac{d^2W}{dx^2}\right) + D_{11}\left(\frac{d^2W}{dx^2}\right)^2\right]dx \tag{7.20}$$

$$T_{max} = \frac{\omega^2}{2}\int_0^L\left[I_0\left(U^2 + W^2\right) - 2I_1\left(U\right)\left(\frac{dW}{dx}\right) + I_2\left(\frac{dW}{dx}\right)^2\right]dx \tag{7.21}$$

Substituting Eq. (7.20) and Eq. (7.21) into the Lagrangian energy function $\Pi = S_{max} - T_{max}$ and setting $\Pi = 0$, one gets:

$$\int_0^L\left[A_{11}\left(\frac{dU}{dx}\right)^2 - 2B_{11}\left(\frac{d^2W}{dx^2}\right)\left(\frac{dU}{dx}\right) + D_{11}\left(\frac{d^2W}{dx^2}\right)^2\right]dx$$
$$= \omega^2\int_0^L\left[I_0\left(U^2 + W^2\right) - 2I_1\left(U\right)\left(\frac{dW}{dx}\right) + I_2\left(\frac{dW}{dx}\right)^2\right]dx \tag{7.22}$$

Where,

$$A_{11} = \frac{h}{1-v^2}\left[\frac{\left(E_c - \tilde{E}_m(\alpha_1,\beta_1)\right)}{k+1} + \tilde{E}_m(\alpha_1,\beta_1)\right], \; B_{11} = \frac{h^2 k}{1-v^2}\left[\frac{\left(E_c - \tilde{E}_m(\alpha_1,\beta_1)\right)}{2(k+1)(k+2)}\right],$$

$$D_{11} = \frac{h^3}{1-v^2}\left[\frac{\left(E_c - \tilde{E}_m(\alpha_1,\beta_1)\right)\left(k^2+k+2\right)}{4(k+1)(k+2)(k+3)} + \frac{\tilde{E}_m(\alpha_1,\beta_1)}{12}\right],$$

$$I_0 = h\left[\frac{\left(\rho_c - \tilde{\rho}_m(\alpha_2,\beta_2)\right)}{k+1} + \tilde{\rho}_m(\alpha_2,\beta_2)\right], \; I_1 = h^2 k\left[\frac{\left(\rho_c - \tilde{\rho}_m(\alpha_2,\beta_2)\right)}{2(k+1)(k+2)}\right]$$

$$I_2 = h^3\left[\frac{\left(\rho_c - \tilde{\rho}_m(\alpha_2,\beta_2)\right)\left(k^2+k+2\right)}{4(k+1)(k+2)(k+3)} + \frac{\tilde{\rho}_m(\alpha_2,\beta_2)}{12}\right]$$

7.3 APPLICATION OF RAYLEIGH–RITZ METHOD

Rayleigh–Ritz method is a powerful numerical technique used to approximate the natural frequencies and mode shapes of vibrating structures. It is particularly useful when dealing with complex structures or boundary conditions where obtaining analytical solutions is challenging. The Rayleigh–Ritz method aims to minimize a suitable functional, such as the Rayleigh quotient or the total potential energy, with respect to the coefficients of the trial functions. This minimization process leads to a system of algebraic equations, known as the generalized eigenvalue problem, which can be solved to determine the natural frequencies and mode shapes of the vibrating structure.

The axial displacement $U(X)$ and transverse displacement $W(X)$ can be now expressed as [3]:

$$U(X) = X^m (R-X)^n \sum_{i=1}^{N} c_i X^{i-1} \tag{7.23a}$$

$$W(X) = X^m (R-X)^n \sum_{i=1}^{N} d_i X^{i-1} \tag{7.23b}$$

Here $c_i's$ and $d_i's$ are unknown coefficients, X^i is the ith term of shape function, and $X^m (R-X)^n$ is the admissible function with exponents m and n. For simply supported–simply supported (SS), clamped–simply supported (CS), and clamped–clamped (CC) boundary conditions, we have $m = n = 1$, $m = 2$ and $n = 2$, and $m = n = 2$, respectively.

Substituting Eq. (7.23a) and Eq. (7.23b) into Eq. (7.22) and differentiating partially with respect to the unknown coefficients $c_i's$ and $d_i's$, $i = 1,2,3...N$, give rise to the generalized eigenvalue problem as:

$$[K]\{X\} = \Omega^2 [M]\{X\} \tag{7.24}$$

where $\{X\} = \left[c_1, c_2, c_3, \ldots c_N, d_1, d_2, d_3, \ldots d_N\right]^T$, $[K]$ represents the stiffness matrix, and $[M]$ denotes the mass matrix.

7.4   NUMERICAL RESULTS AND DISCUSSIONS

In this chapter, the FG beam is composed of metal constituents as aluminum (Al) and ceramic constituent as alumina (Al$_2$O$_3$). The mechanical properties are given as [1–4]:

> Alumina (Al$_2$O$_3$): $E_c = 380\,\mathrm{GPa}$, $\rho_c = 3800\ kg.m^{-3}$, and $v_c = 0.3$
> Aluminum (Al): $E_m = (60, 70, 80)\,\mathrm{GPa}$, $\rho_m = (2690, 2700, 2710)\ kg.m^{-3}$, and
> $v_m = 0.3$.

In this context, the Young's modulus and material density of aluminum (Al) are represented using Triangular Fuzzy Numbers. Furthermore, variations in Young's modulus and mass density across the thickness are modeled on the basis of the power-law exponent model.

7.4.1   VALIDATION AND CONVERGENCE

In this subsection, a comparison has been made between the current uncertain model and existing deterministic models in specific scenarios. Specifically, the natural frequencies of functionally graded beams under SS, CS, and CC boundary conditions were compared, with a focus on scenarios where uncertainties in material properties were disregarded. The numerical computations for SS, CS, and CC boundary conditions were carried out using the Rayleigh–Ritz method. Additionally, the first, second, and third modes of frequency parameters $\left(\Omega = \dfrac{\omega L^2}{h} \sqrt{\dfrac{\rho_m}{E_m}}\right)$ for SS, CS, and CC boundary conditions were compared with the existing results [1]. Here, the materials alumina (Al$_2$O$_3$) and aluminum (Al) are considered, and the gradation is carried out along Young's modulus only, incorporating $E_m = 70\,\mathrm{GPa}$, $E_c = 380$ GPa, and $v = 0.3$. Table 7.1(a, b, and c) presents the tabular results for first three modes of frequency parameters for different power-law exponents with aspect ratio of $L/h = 20$. These findings illustrate the accuracy and robustness of the current model, showing its capability to handle various scenarios effectively and aligning well with existing results in specific cases.

   To demonstrate the accuracy, efficiency, and powerfulness of the present analysis, convergence study is carried out. For test case, simply supported–simply supported (SS) boundary condition is taken for investigations. First four modes of frequency parameters for lower and bounds are computed for different values of N with $\tilde{E}_m = (60, 70, 80)\,\mathrm{GPa}$, $\tilde{\rho}_L = (2690, 2700, 2710)\ kg.m^{-3}$, $L/h = 10$, $k = 5$, and $\alpha_1 = \alpha_2 = 0.5$, which are demonstrated in Table 7.2. The results indicate that the first-mode frequency converges more rapidly with six terms compared to higher modes. However, as the number of terms increases to 16, all mode frequencies demonstrate convergence.

Table 7.1
Comparison of Present Results with Existing Results in the Special Case:
(a) Frequency Parameter of SS Boundary Condition with $L/h = 20$

(k)		0	0.2	0.5	1	2	5	10
Ω_1	Present	6.9516	6.3355	5.7627	5.2563	4.8259	4.3803	4.0208
	[1]	6.9516	6.3355	5.7627	5.2563	4.8259	4.3803	4.0208
Ω_2	Present	27.7212	25.0958	22.3005	19.5324	17.2307	15.8432	15.1403
	[1]	27.7212	25.0558	22.3005	19.5325	17.2307	15.8432	15.1403
Ω_3	Present	62.0572	56.2196	50.0855	44.0796	39.0859	35.8720	34.1082
	[1]	62.0573	56.2196	50.0856	44.0796	39.0854	35.8721	34.1082

(b) Frequency Parameters of CS Boundary Condition with $L/h = 20$

(k)		0	0.2	0.5	1	2	5	10
Ω_1	Present	10.8579	9.8432	8.7902	7.7689	6.9195	6.3447	6.0063
	[1]	10.8579	9.8444	8.7944	7.7777	6.9322	6.3550	6.0119
Ω_2	Present	35.0723	31.7650	28.2723	24.8365	21.9800	20.1918	19.2356
	[1]	35.0723	31.7671	28.2800	24.8426	22.0035	20.2106	19.2458
Ω_3	Present	72.7922	65.9101	58.6074	51.3953	45.3934	41.7101	39.8112
	[1]	72.7922	65.9167	58.6315	51.4455	45.4665	41.7697	39.8443

(c) Frequency Parameters of CC Boundary Condition with $L/h = 20$

(k)		0	0.2	0.5	1	2	5	10
Ω_1	Present	15.7545	14.2669	12.6913	11.1363	9.8433	9.0496	8.6313
	[1]	15.7545	14.2688	12.6986	11.1615	9.8641	9.0663	8.6408
Ω_2	Present	43.2764	39.1884	34.8552	30.5759	27.0159	24.8364	23.6968
	[1]	43.2769	39.1943	34.8785	30.6220	27.0776	24.8867	23.7258
Ω_3	Present	84.3802	76.4020	67.9321	59.5597	52.5931	48.3475	46.1564
	[1]	84.3802	76.4118	67.9712	59.6405	52.7041	48.4365	46.2069

Table 7.2
Frequency Parameters Ω for Different Values of N with $\tilde{E}_m = (60,70,80)\,\text{GPa}$, $\tilde{\rho}_L = (2690,2700,2710)\,\text{Kg.m}^{-3}$, $\dfrac{L}{h} = 10$, $k = 5$, with $\alpha_1 = \alpha_2 = 0.5$ for SS Boundary Condition

N	$\beta_1 = 0$ (Lower Bound)				$\beta_1 = 1$ (Upper Bound)			
	$\underline{\Omega}_1$	$\underline{\Omega}_2$	$\underline{\Omega}_3$	$\underline{\Omega}_4$	$\bar{\Omega}_1$	$\bar{\Omega}_2$	$\bar{\Omega}_3$	$\bar{\Omega}_4$
2	4.9312	18.8642	44.6660	88.3731	4.6972	5.8550	18.2591	18.5188
4	4.3229	15.2140	41.4618	48.9064	4.1405	6.2685	14.8864	19.4686
6	4.3212	15.1544	34.4284	41.0600	4.1389	7.1874	14.8274	22.2444
8	4.3212	15.1541	33.8694	41.0276	4.1389	8.9836	14.8271	27.7530
10	4.3212	15.1541	33.8611	41.0271	4.1389	12.6851	14.8271	33.0450
12	4.3212	15.1541	33.8610	41.0271	4.1389	14.8271	21.9934	33.0450
14	4.3212	15.1541	33.8610	41.0271	4.1389	14.8271	33.0450	39.9011
16	4.3212	15.1541	33.8610	41.0271	4.1389	14.8271	33.0450	39.9011

7.4.2 Propagation of Uncertainties

In this subsection, three cases are analyzed in order to study the propagation of uncertainty in the frequency parameters of the FG beam. In the first case, Young's modulus (E_m) is taken as fuzzy or imprecise, while in the second and third cases, the mass density (ρ_m) and both parameters (Young's modulus (E_m) and mass density(ρ_m)) of metal constituent are taken as imprecise or fuzzy input parameters. The uncertainties in frequency parameters of FG beam for both the lower and upper bounds of SS, CS, and CC boundary conditions for the first case are shown in Table 7.3(a, b,

Table 7.3

Lower Bound and Upper Bound of Frequency Parameters (Ω) for Different Values of α_1 with $\tilde{E}_m = (60,70,80)\text{GPa}$, $\dfrac{L}{h} = 10$, and $k = 5$

(a) SS Boundary Condition

α_1	$\beta_1 = 0$ (Lower Bound)				$\beta_1 = 1$ (Upper Bound)			
	$\underline{\Omega}_1$	$\underline{\Omega}_2$	$\underline{\Omega}_3$	$\underline{\Omega}_4$	$\bar{\Omega}_1$	$\bar{\Omega}_2$	$\bar{\Omega}_3$	$\bar{\Omega}_4$
0	4.4311	15.3423	34.3367	41.7037	4.0596	14.6733	32.6683	39.4006
0.1	4.4082	15.3041	34.2395	41.5641	4.0747	14.7027	32.7403	39.4954
0.2	4.3859	15.2666	34.1441	41.4276	4.0901	14.7325	32.8133	39.5920
0.3	4.3641	15.2295	34.0503	41.2940	4.1058	14.7627	32.8873	39.6904
0.4	4.3429	15.1931	33.9582	41.1634	4.1218	14.7932	32.9624	39.7907
0.5	4.3221	15.1572	33.8677	41.0355	4.1381	14.8242	33.0385	39.8929
0.6	4.3018	15.1217	33.7786	40.9104	4.1548	14.8555	33.1158	39.9971
0.7	4.2820	15.0868	33.6911	40.7879	4.1718	14.8872	33.1942	40.1034
0.8	4.2626	15.0524	33.6050	40.6679	4.1892	14.9194	33.2739	40.2118
0.9	4.2436	15.0185	33.5202	40.5503	4.2069	14.9520	33.3547	40.3224
1	4.2251	14.9850	33.4368	40.4352	4.2251	14.9850	33.4368	40.4352

(b) CS Boundary Condition

α_1	$\beta_1 = 0$ (Lower Bound)				$\beta_1 = 1$ (Upper Bound)			
	$\underline{\Omega}_1$	$\underline{\Omega}_2$	$\underline{\Omega}_3$	$\underline{\Omega}_4$	$\bar{\Omega}_1$	$\bar{\Omega}_2$	$\bar{\Omega}_3$	$\bar{\Omega}_4$
0	6.3091	19.6668	39.0021	44.2961	5.9525	18.7183	37.3191	41.2170
0.1	6.2880	19.6118	38.9039	44.1061	5.9676	18.7594	37.3914	41.3418
0.2	6.2673	19.5577	38.8073	43.9207	5.9830	18.8010	37.4647	41.4690
0.3	6.2470	19.5046	38.7125	43.7399	5.9986	18.8432	37.5391	41.5988
0.4	6.2271	19.4524	38.6194	43.5634	6.0144	18.8861	37.6146	41.7313
0.5	6.2076	19.4010	38.5279	43.3910	6.0305	18.9295	37.6913	41.8665
0.6	6.1884	19.3505	38.4379	43.2227	6.0469	18.9735	37.7691	42.0045
0.7	6.1696	19.3008	38.3495	43.0583	6.0635	19.0182	37.8481	42.1454
0.8	6.1511	19.2519	38.2625	42.8975	6.0804	19.0635	37.9284	42.2894
0.9	6.1330	19.2037	38.1769	42.7404	6.0976	19.1095	38.0099	42.4365
1	6.1151	19.1562	38.0927	42.5868	6.1151	19.1562	38.0927	42.5868

(Continued)

Table 7.3 (Continued)

Lower Bound and Upper Bound of Frequency Parameters (Ω) for Different Values of α_1 with $\tilde{E}_m = (60,70,80)$ GPa, $\dfrac{L}{h} = 10$, and $k = 5$

(c) CC Boundary Condition

α_1	$\beta_1 = 0$ (Lower Bound)				$\beta_1 = 1$ (Upper Bound)			
	$\underline{\Omega}_1$	$\underline{\Omega}_2$	$\underline{\Omega}_3$	$\underline{\Omega}_4$	$\overline{\Omega}_1$	$\overline{\Omega}_2$	$\overline{\Omega}_3$	$\overline{\Omega}_4$
0	8.9627	24.2562	44.0348	46.2895	8.5191	23.0894	40.8492	44.1524
0.1	8.9370	24.1888	43.8331	46.1669	8.5383	23.1401	40.9747	44.2458
0.2	8.9118	24.1226	43.6370	46.0462	8.5578	23.1915	41.1031	44.3405
0.3	8.8869	24.0575	43.4462	45.9275	8.5775	23.2436	41.2344	44.4363
0.4	8.8625	23.9934	43.2606	45.8105	8.5976	23.2964	41.3686	44.5335
0.5	8.8385	23.9303	43.0800	45.6954	8.6179	23.3499	41.5061	44.6319
0.6	8.8149	23.8683	42.9041	45.5820	8.6385	23.4042	41.6467	44.7317
0.7	8.7916	23.8072	42.7327	45.4703	8.6594	23.4593	41.7907	44.8329
0.8	8.7687	23.7470	42.5657	45.3602	8.6806	23.5152	41.9382	44.9354
0.9	8.7462	23.6877	42.4029	45.2517	8.7021	23.5718	42.0893	45.0393
1	8.7240	23.6294	42.2442	45.1448	8.7240	23.6294	42.2442	45.1448

and c), whereas Figure 7.1(a, b, and c) is the graphical representation of fundamental frequency parameters. Likewise, Table 7.4(a, b, and c) displays the propagation of uncertainty in the frequency parameters for all the four modes of frequency parameters while Figure 7.2(a, b, and c) illustrates the graphical results for the second case with respect to fundamental mode. Table 7.5(a, b, and c) shows the propagation of uncertainty for the third case with Figure 7.3(a, b, and c) as graphical results. It is essential to note that the frequency parameters are obtained for the deterministic (or exact or crisp) case by considering $\alpha_1 = 1$ or/and $\alpha_2 = 1$. It should also be noted that $\beta = 0$ gives the lower bound, while $\beta = 1$ gives the upper bound of the frequency parameters. It is also important to note that by taking $\alpha = 1$, both the upper and lower bounds give the same results, which are the same as deterministic (or crisp) value. The computations are carried out in all of the above cases by considering the aspect ratio $L/h = 10$ and power-law exponent $k = 5$. From Figure 7.4, it is concluded that the spreads or fuzziness of Triangular fuzzy output of frequencies is the highest in case of uncertain Young's modulus (E_m), whereas the fuzziness in frequencies is the lowest in case of uncertain mass density (ρ_m) as input. Also, the fuzziness of both the Young's modulus and mass density as input is matching with uncertain Young's modulus as input, and the same pattern is observed in all the three boundary conditions.

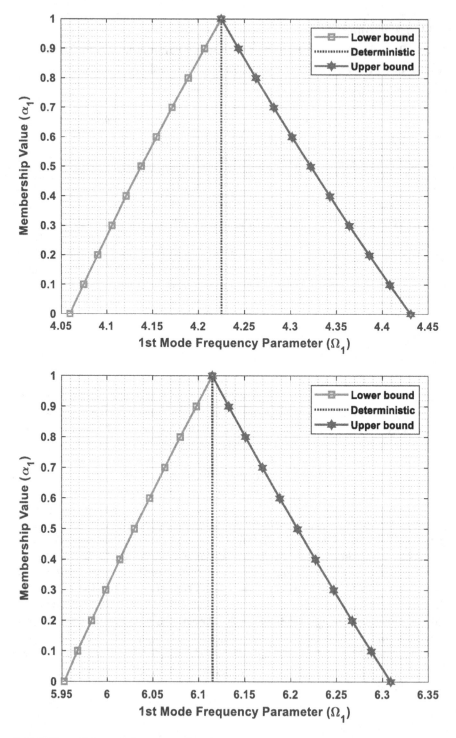

FIGURE 7.1   Triangular Fuzzy Number (TFN) when E_m is fuzzy for (a) SS boundary condition, (b) CS boundary condition, and (c) CC boundary condition.

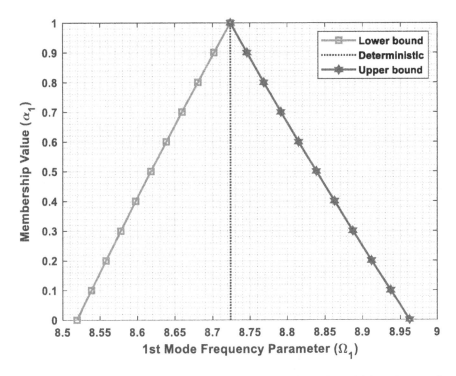

FIGURE 7.1 Triangular Fuzzy Number (TFN) when E_m is fuzzy for (a) SS boundary condition, (b) CS boundary condition, and (c) CC boundary condition.

Table 7.4
Lower Bound and Upper Bound of Frequency Parameters (Ω) for Different Values of α_2 with $\tilde{\rho}_L = (2690, 2700, 2710)\ \text{kg.m}^{-3}$, $\dfrac{L}{h} = 10$, and $k = 5$
(a) SS Boundary Condition

	$\beta_2 = 0$ (Lower Bound)				$\beta_2 = 1$ (Upper Bound)			
α_2	$\underline{\Omega}_1$	$\underline{\Omega}_2$	$\underline{\Omega}_3$	$\underline{\Omega}_4$	$\bar{\Omega}_1$	$\bar{\Omega}_2$	$\bar{\Omega}_3$	$\bar{\Omega}_4$
0	4.2233	14.9790	33.4237	40.4186	4.2268	14.9910	33.4499	40.4517
0.1	4.2235	14.9796	33.4250	40.4203	4.2266	14.9904	33.4486	40.4501
0.2	4.2237	14.9802	33.4263	40.4219	4.2264	14.9898	33.4473	40.4484
0.3	4.2239	14.9808	33.4276	40.4236	4.2263	14.9892	33.4460	40.4468
0.4	4.2240	14.9814	33.4289	40.4253	4.2261	14.9886	33.4446	40.4451
0.5	4.2242	14.9820	33.4303	40.4269	4.2259	14.9880	33.4433	40.4435
0.6	4.2244	14.9826	33.4316	40.4286	4.2258	14.9874	33.4420	40.4418
0.7	4.2246	14.9832	33.4329	40.4302	4.2256	14.9868	33.4407	40.4402
0.8	4.2247	14.9838	33.4342	40.4319	4.2254	14.9862	33.4394	40.4385
0.9	4.2249	14.9844	33.4355	40.4336	4.2252	14.9856	33.4381	40.4369
1	4.2251	14.9850	33.4368	40.4352	4.2251	14.9850	33.4368	40.4352

Table 7.4 (*Continued*)

Lower Bound and Upper Bound of Frequency Parameters (Ω) for Different Values of α_2 with $\tilde{\rho}_l = (2690, 2700, 2710)\ \text{kg.m}^{-3}$, $\dfrac{L}{h} = 10$, and $k = 5$

(b) CS Boundary Condition

α_2	$\underline{\Omega}_1$	$\underline{\Omega}_2$	$\underline{\Omega}_3$	$\underline{\Omega}_4$	$\bar{\Omega}_1$	$\bar{\Omega}_2$	$\bar{\Omega}_3$	$\bar{\Omega}_4$
	$\beta_2 = 0$ (Lower Bound)				$\beta_2 = 1$ (Upper Bound)			
0	6.1127	19.1486	38.0776	42.5695	6.1176	19.1638	38.1078	42.6039
0.1	6.1129	19.1494	38.0791	42.5712	6.1174	19.1631	38.1063	42.6022
0.2	6.1132	19.1502	38.0807	42.5730	6.1171	19.1623	38.1048	42.6005
0.3	6.1134	19.1509	38.0822	42.5747	6.1169	19.1616	38.1033	42.5988
0.4	6.1137	19.1517	38.0837	42.5764	6.1166	19.1608	38.1018	42.5971
0.5	6.1139	19.1524	38.0852	42.5781	6.1164	19.1600	38.1003	42.5953
0.6	6.1141	19.1532	38.0867	42.5799	6.1161	19.1593	38.0988	42.5936
0.7	6.1144	19.1540	38.0882	42.5816	6.1159	19.1585	38.0973	42.5919
0.8	6.1146	19.1547	38.0897	42.5833	6.1156	19.1578	38.0958	42.5902
0.9	6.1149	19.1555	38.0912	42.5850	6.1154	19.1570	38.0943	42.5885
1	6.1151	19.1562	38.0927	42.5868	6.1151	19.1562	38.0927	42.5868

(c) CC Boundary Condition

α_2	$\underline{\Omega}_1$	$\underline{\Omega}_2$	$\underline{\Omega}_3$	$\underline{\Omega}_4$	$\bar{\Omega}_1$	$\bar{\Omega}_2$	$\bar{\Omega}_3$	$\bar{\Omega}_4$
	$\beta_2 = 0$ (Lower Bound)				$\beta_2 = 1$ (Upper Bound)			
0	8.7205	23.6199	42.2268	45.1273	8.7275	23.6387	42.2614	45.1622
0.1	8.7208	23.6209	42.2286	45.1290	8.7272	23.6378	42.2597	45.1604
0.2	8.7212	23.6218	42.2303	45.1308	8.7268	23.6368	42.2580	45.1587
0.3	8.7215	23.6228	42.2321	45.1325	8.7265	23.6359	42.2563	45.1570
0.4	8.7219	23.6237	42.2338	45.1343	8.7261	23.6350	42.2545	45.1552
0.5	8.7222	23.6247	42.2355	45.1360	8.7258	23.6340	42.2528	45.1535
0.6	8.7226	23.6256	42.2373	45.1378	8.7254	23.6331	42.2511	45.1517
0.7	8.7229	23.6265	42.2390	45.1395	8.7250	23.6322	42.2494	45.1500
0.8	8.7233	23.6275	42.2407	45.1413	8.7247	23.6312	42.2476	45.1483
0.9	8.7236	23.6284	42.2425	45.1430	8.7243	23.6303	42.2459	45.1465
1	8.7240	23.6294	42.2442	45.1448	8.7240	23.6294	42.2442	45.1448

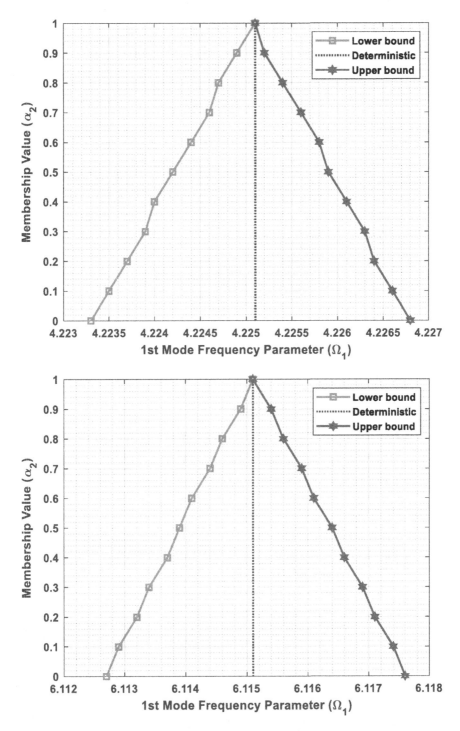

FIGURE 7.2 Triangular Fuzzy Number (TFN) when ρ_m is fuzzy for (a) SS boundary condition, (b) CS boundary condition, and (c) CC boundary condition.

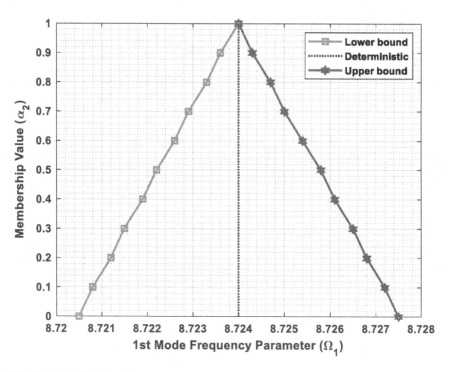

FIGURE 7.2 (Continued)

Table 7.5

Lower Bound and Upper Bound of Frequency Parameters (Ω) for Different Values of α_1 and α_2 with $\tilde{E}_m = (60,70,80)\,\text{GPa}$, $\tilde{\rho}_L = (2690,2700,2710)\,\text{Kg.m}^{-3}$, $\frac{L}{h} = 10$, and $k = 5$

(a) SS Boundary Condition

α_1	$\beta_1 = \beta_2 = 0$ (Lower Bound)				$\beta_1 = \beta_2 = 1$ (Upper Bound)			
α_2	$\underline{\Omega}_1$	$\underline{\Omega}_2$	$\underline{\Omega}_3$	$\underline{\Omega}_4$	$\bar{\Omega}_1$	$\bar{\Omega}_2$	$\bar{\Omega}_3$	$\bar{\Omega}_4$
0	4.4293	15.3362	34.3233	41.6866	4.0613	14.6791	32.6812	39.4167
0.1	4.4066	15.2987	34.2275	41.5488	4.0762	14.7080	32.7519	39.5099
0.2	4.3845	15.2617	34.1334	41.4140	4.0914	14.7372	32.8236	39.6049
0.3	4.3629	15.2253	34.0410	41.2822	4.1069	14.7668	32.8963	39.7017
0.4	4.3418	15.1895	33.9502	41.1533	4.1228	14.7968	32.9701	39.8004
0.5	4.3212	15.1541	33.8610	41.0271	4.1389	14.8271	33.0450	39.9011
0.6	4.3011	15.1193	33.7734	40.9037	4.1554	14.8579	33.1210	40.0037
0.7	4.2814	15.0850	33.6871	40.7829	4.1723	14.8890	33.1981	40.1083
0.8	4.2622	15.0512	33.6023	40.6645	4.1895	14.9206	33.2765	40.2151
0.9	4.2434	15.0179	33.5189	40.5487	4.2071	14.9526	33.3560	40.3240
1	4.2251	14.9850	33.4368	40.4352	4.2251	14.9850	33.4368	40.4352

(Continued)

Table 7.5 (*Continued*)

Lower Bound and Upper Bound of Frequency Parameters (Ω) for Different Values of α_1 and α_2 with $\tilde{E}_m = (60, 70, 80)$ GPa, $\tilde{\rho}_L = (2690, 2700, 2710)$ Kg.m^{-3}, $\frac{L}{h} = 10$, and $k = 5$

(b) CS Boundary Condition

α_2	$\beta_2 = 0$ (Lower Bound)				$\beta_2 = 1$ (Upper Bound)			
	$\underline{\Omega}_1$	$\underline{\Omega}_2$	$\underline{\Omega}_3$	$\underline{\Omega}_4$	$\bar{\Omega}_1$	$\bar{\Omega}_2$	$\bar{\Omega}_3$	$\bar{\Omega}_4$
0	6.3066	19.6590	38.9868	44.2782	5.9549	18.7258	37.3340	41.2336
0.1	6.2857	19.6048	38.8901	44.0900	5.9698	18.7661	37.4048	41.3568
0.2	6.2653	19.5515	38.7951	43.9065	5.9849	18.8070	37.4766	41.4824
0.3	6.2452	19.4992	38.7019	43.7275	6.0003	18.8485	37.5496	41.6106
0.4	6.2256	19.4478	38.6103	43.5528	6.0159	18.8906	37.6236	41.7414
0.5	6.2063	19.3972	38.5203	43.3823	6.0317	18.9332	37.6988	41.8749
0.6	6.1874	19.3474	38.4318	43.2157	6.0478	18.9765	37.7751	42.0113
0.7	6.1688	19.2985	38.3449	43.0530	6.0642	19.0204	37.8526	42.1505
0.8	6.1506	19.2503	38.2595	42.8941	6.0809	19.0650	37.9314	42.2928
0.9	6.1327	19.2029	38.1754	42.7387	6.0979	19.1103	38.0114	42.4382
1	6.1151	19.1562	38.0927	42.5868	6.1151	19.1562	38.0927	42.5868

(c) CC Boundary Condition

α_1 α_2	$\beta_1 = \beta_2 = 0$ (Lower Bound)				$\beta_1 = \beta_2 = 1$ (Upper Bound)			
	$\underline{\Omega}_1$	$\underline{\Omega}_2$	$\underline{\Omega}_3$	$\underline{\Omega}_4$	$\bar{\Omega}_1$	$\bar{\Omega}_2$	$\bar{\Omega}_3$	$\bar{\Omega}_4$
0	8.9591	24.2465	44.0168	46.2718	8.5226	23.0986	40.8658	44.1695
0.1	8.9338	24.1802	43.8169	46.1510	8.5414	23.1484	40.9898	44.2613
0.2	8.9089	24.1149	43.6227	46.0321	8.5606	23.1989	41.1165	44.3542
0.3	8.8844	24.0508	43.4338	45.9151	8.5800	23.2501	41.2461	44.4484
0.4	8.8604	23.9877	43.2500	45.8000	8.5997	23.3020	41.3788	44.5439
0.5	8.8367	23.9256	43.0712	45.6866	8.6196	23.3546	41.5145	44.6406
0.6	8.8134	23.8645	42.8970	45.5750	8.6399	23.4080	41.6535	44.7387
0.7	8.7905	23.8043	42.7275	45.4650	8.6604	23.4621	41.7959	44.8381
0.8	8.7680	23.7451	42.5622	45.3567	8.6813	23.5170	41.9417	44.9389
0.9	8.7458	23.6868	42.4012	45.2500	8.7025	23.5728	42.0911	45.0411
1	8.7240	23.6294	42.2442	45.1448	8.7240	23.6294	42.2442	45.1448

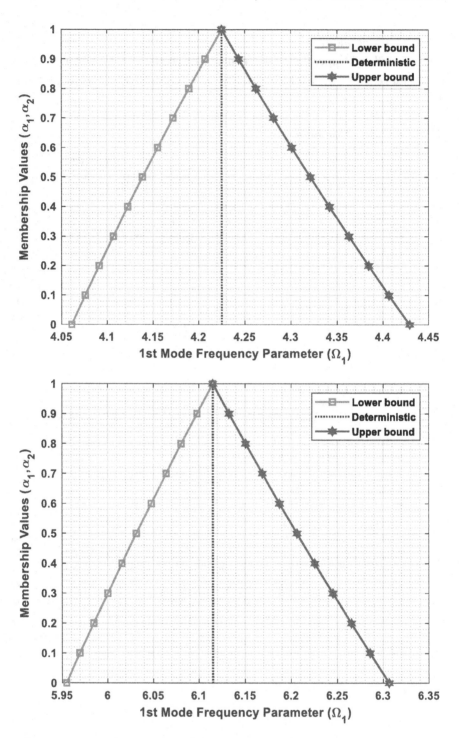

FIGURE 7.3   Triangular Fuzzy Number (TFN) when E_m and ρ_m are fuzzy for (a) SS boundary condition, (b) CS boundary condition, and (c) CC boundary condition.

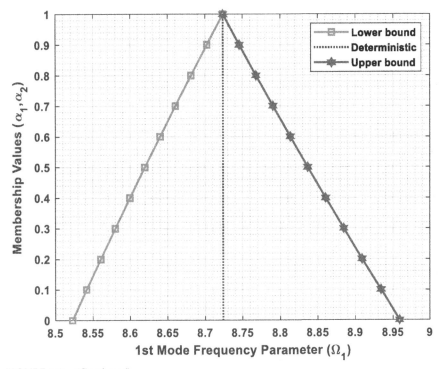

FIGURE 7.3   (Continued)

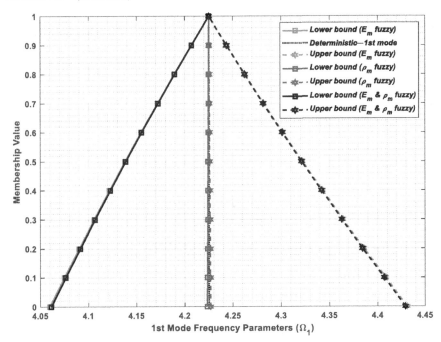

FIGURE 7.4   Comparisons of Triangular Fuzzy Number (TFN) for HH boundary condition.

7.5 CONCLUSION

In this chapter, Rayleigh–Ritz method is employed to calculate the frequency parameters of the uncertain model for simply supported–simply supported (SS), clamped–simply supported (CS), and clamped–clamped (CC) boundary conditions of the FG beam as lower and upper bounds. Here, the uncertainties are assumed to be associated with Young's modulus and material density of the metal constituent as Triangular Fuzzy Number. The results obtained by the uncertain model are validated with the deterministic model exhibiting a robust agreement. Further, a parametric study is conducted to investigate the fuzziness or spreads of the natural frequency concerning different uncertain parameters. In the case of uncertain Young's modulus, the spreads or fuzziness of triangular fuzzy frequency output is the highest, whereas in the case of uncertain mass density as input, the fuzziness in frequencies is the lowest. Also, the fuzziness of both the Young's modulus and the mass density as input is consistent with the uncertain Young's modulus as input.

BIBLIOGRAPHY

[1] Pradhan, K.K. and Chakraverty, S., 2013. Free vibration of Euler and Timoshenko functionally graded beams by Rayleigh—Ritz method. *Composites Part B: Engineering, 51*, pp. 175–184.

[2] Pradhan, K.K. and Chakraverty, S., 2014. Effects of different shear deformation theories on free vibration of functionally graded beams. *International Journal of Mechanical Sciences, 82*, pp. 149–160.

[3] Chakraverty, S. and Pradhan, K.K., 2016. *Vibration of functionally graded beams and plates.* Academic Press.

[4] Aydogdu, M. and Taskin, V., 2007. Free vibration analysis of functionally graded beams with simply supported edges. *Materials & Design, 28*(5), pp. 1651–1656.

[5] Wattanasakulpong, N. and Ungbhakorn, V., 2014. Linear and nonlinear vibration analysis of elastically restrained ends FGM beams with porosities. *Aerospace Science and Technology, 32*(1), pp. 111–120.

[6] Reddy, J.N., 2007. Nonlocal theories for bending, buckling and vibration of beams. *International Journal of Engineering Science, 45*(2–8), pp. 288–307.

[7] Jena, S.K., Chakraverty, S., Malikan, M. and Sedighi, H., 2020. Implementation of Hermite—Ritz method and Navier's technique for vibration of functionally graded porous nanobeam embedded in Winkler—Pasternak elastic foundation using bi-Helmholtz nonlocal elasticity. *Journal of Mechanics of Materials and Structures, 15*(3), pp. 405–434.

[8] Jena, S.K., Chakraverty, S. and Malikan, M., 2020. Application of shifted Chebyshev polynomial-based Rayleigh—Ritz method and Navier's technique for vibration analysis of a functionally graded porous beam embedded in Kerr foundation. *Engineering with Computers*, pp. 1–21.

8 Vibration of Functionally Graded Beam with Complicating Effects in Uncertain Environment

8.1 INTRODUCTION

This chapter seeks to investigate the effect of geometrical uncertainties on the free vibration of Euler–Bernoulli functionally graded (FG) beams resting on Winkler–Pasternak elastic foundation. In this scenario, the uncertainties are linked to length and thickness of FG beam using Symmetric Gaussian Fuzzy Number (SGFN). The governing equations of motion regarding free vibration of the uncertain model are derived by combining the Symmetric Gaussian Fuzzy Number with Hamilton's principle and double parametric form of fuzzy numbers. The natural frequencies of the uncertain models are computed using the double parametric form-based Navier's approach for hinged–hinged (H-H) boundary condition. The double parametric form-based Hermite–Ritz approach was also used to calculate the natural frequencies of hinged–hinged (H-H), clamped–hinged (C-H), and clamped–clamped (C-C) boundary conditions. Natural frequencies obtained using Navier's method and Hermite–Ritz method are used to validate the results of the uncertain model which exhibit a strong agreement. A comprehensive parametric analysis is also conducted with respect to various graphical and tabular results to investigate the fuzziness or spreads of natural frequencies in relation to various uncertain parameters.

8.2 MATHEMATICAL FORMULATION OF THE PROPOSED PROBLEM

As per power-law model, Young's modulus $E(z)$ and material density $\rho(z)$ of the FG beam can be expressed as [1–4]:

$$E(z) = (E_c - E_m)\left(\frac{z}{h} + \frac{1}{2}\right)^k + E_m \tag{8.1}$$

$$\rho(z) = (\rho_c - \rho_m)\left(\frac{z}{h} + \frac{1}{2}\right)^k + \rho_m \tag{8.2}$$

Here, E_c and E_m symbolize Young's modulus for ceramic and metal constituents, respectively. Likewise, ρ_c and ρ_m denote the material density for ceramic and metal constituents, respectively. k is the power-law exponent that regulates material

DOI: 10.1201/9781003303107-8

119

distribution along the thickness of the beam, z is the distance from the mid-plane of the FG beam, and h is the thickness of the FG beam.

The displacement field according to the Euler–Bernoulli beam theory can be given as [5]:

$$u_1(x,z,t) = u(x,t) - z\left(\frac{\partial w}{\partial x}\right)$$

$$u_2(x,z,t) = 0 \tag{8.3}$$

$$u_3(x,z,t) = w(x,t)$$

Axial and transverse displacements on the mid-plane of the FG beam are represented by $u(x,t)$ and $w(x,t)$, respectively.

The strain–displacement relation of the FG beam is stated as:

$$\varepsilon_{xx} = \frac{\partial u_1(x,z,t)}{\partial x} = \frac{\partial u(x,t)}{\partial x} - z\frac{\partial^2 w(x,t)}{\partial x^2} \tag{8.4}$$

The stress component of the FG beam by the generalized Hooke's law[4]:

$$\sigma_{xx} = Q_{11}\varepsilon_{xx} = \left(\frac{E(z)}{1-v^2}\right)\varepsilon_{xx} \tag{8.5}$$

The strain energy (S) of the FG beam is stated as [1]:

$$
\begin{aligned}
S &= \frac{1}{2}\int_0^L \int_A (\sigma_{xx}\varepsilon_{xx})\,dA\,dx \\
&= \frac{1}{2}\int_0^L \int_A \left[\sigma_{xx}\left(\frac{\partial u(x,t)}{\partial x} - z\frac{\partial^2 w(x,t)}{\partial x^2}\right)\right]dA\,dx \\
&= \frac{1}{2}\int_0^L \left[N\left(\frac{\partial u(x,t)}{\partial x}\right) - M\left(\frac{\partial^2 w(x,t)}{\partial x^2}\right)\right]dx
\end{aligned}
\tag{8.6}
$$

where the stress resultants $(N,M) = \int_A (\sigma_{xx}, z\sigma_{xx})\,dA$.

Now, the variation in strain energy (δS) can be written as [1]:

$$
\begin{aligned}
\delta S &= \int_0^L \int_A (\sigma_{xx}\delta\varepsilon_{xx})\,dA\,dx \\
&= \int_0^L \int_A \left[\sigma_{xx}\left(\frac{\partial\delta u(x,t)}{\partial x} - z\frac{\partial^2\delta w(x,t)}{\partial x^2}\right)\right]dA\,dx \\
&= \int_0^L \left[N\left(\frac{\partial\delta u(x,t)}{\partial x}\right) - M\left(\frac{\partial^2\delta w(x,t)}{\partial x^2}\right)\right]dx \\
&= \int_0^L \left[-\left(\frac{\partial N}{\partial x}\right)\delta u - \left(\frac{\partial^2 M}{\partial x^2}\right)\delta w\right]dx
\end{aligned}
\tag{8.7}
$$

The kinetic energy (T) of the FG beam can be stated as [1]:

$$
\begin{aligned}
T &= \frac{1}{2}\int_0^L \int_A \rho(z)\left[\left(\frac{\partial u_1}{\partial t}\right)^2 + \left(\frac{\partial u_2}{\partial t}\right)^2 + \left(\frac{\partial u_3}{\partial t}\right)^2\right] dA\, dx \\
&= \frac{1}{2}\int_0^L \int_A \rho(z)\left[\left(\frac{\partial u}{\partial t} - z\left(\frac{\partial^2 w}{\partial x \partial t}\right)\right)^2 + \left(\frac{\partial w}{\partial t}\right)^2\right] dA\, dx \\
&= \frac{1}{2}\int_0^L \left[I_0\left[\left(\frac{\partial u}{\partial t}\right)^2 + \left(\frac{\partial w}{\partial t}\right)^2\right] - 2I_1\left(\frac{\partial u}{\partial t}\right)\left(\frac{\partial^2 w}{\partial x \partial t}\right) + I_2\left(\frac{\partial^2 w}{\partial x \partial t}\right)^2\right] dx
\end{aligned}
\tag{8.8}
$$

In which, $(I_0, I_1, I_2) = \int_A \rho(z)\left(1, z, z^2\right) dA$ are the mass moments of inertia.

The variation in kinetic energy (δT) can be obtained from Eq. (8.8) as [1]:

$$
\begin{aligned}
\delta T &= \frac{1}{2}\int_0^L \left[I_0\delta\left[\left(\frac{\partial u}{\partial t}\right)^2 + \left(\frac{\partial w}{\partial t}\right)^2\right] - 2I_1\delta\left(\frac{\partial u}{\partial t}\right)\left(\frac{\partial^2 w_b}{\partial x \partial t}\right) + I_2\delta\left(\frac{\partial^2 w}{\partial x \partial t}\right)^2\right] dx \\
&= \int_0^L \left[\begin{aligned}&-I_0\left(\frac{\partial^2 u}{\partial t^2}\right)\delta u - I_0\left(\frac{\partial^2 w}{\partial t^2}\right)\delta w + I_1\left(\frac{\partial^3 w}{\partial x \partial t^2}\right)\delta u - I_1\left(\frac{\partial^3 u}{\partial x \partial t^2}\right)\delta w \\ &+I_2\left(\frac{\partial^4 w}{\partial x^2 \partial t^2}\right)\delta w\end{aligned}\right] dx
\end{aligned}
\tag{8.9}
$$

The work done (W) by the Winkler–Pasternak elastic foundation can be expressed as [6]:

$$
W = -\frac{1}{2}\int_0^L \left[k_w w^2 + k_g\left(\frac{\partial w}{\partial x}\right)^2\right] dx,
\tag{8.10}
$$

where k_w and k_g are Winkler and Pasternak elastic constants, respectively.

The variation in external work done (δW) can be derived from Eq. (8.10) as:

$$
\delta W = -\int_0^L \left[k_w w + k_g\left(\frac{\partial^2 w}{\partial x^2}\right)\right]\delta w\, dx
\tag{8.11}
$$

Using Eq. (8.7), Eq. (8.9), and Eq. (8.11) in the extended Hamilton's principle $\int_0^T \delta(T - S + W)\, dt = 0$ and collecting the coefficient of δu and δw, the governing equations of motion in terms of stress resultants and displacements can be obtained as:

$$
\frac{\partial N}{\partial x} = I_0\left(\frac{\partial^2 u}{\partial t^2}\right) - I_1\left(\frac{\partial^3 w}{\partial x \partial t^2}\right)
\tag{8.12a}
$$

$$\frac{\partial^2 M}{\partial x^2} = I_0 \left(\frac{\partial^2 w}{\partial t^2} \right) + I_1 \left(\frac{\partial^3 u}{\partial x \partial t^2} \right) - I_2 \left(\frac{\partial^4 w}{\partial x^2 \partial t^2} \right) + k_w w - k_g \left(\frac{\partial^2 w}{\partial x^2} \right) \qquad (8.12b)$$

Multiplying Eq. (8.5) by dA and $z \, dA$ and integrating over the area of cross-section of the FG beam, the local stress resultants can be written as [1]:

$$N = A_{11} \left(\frac{\partial u}{\partial x} \right) - B_{11} \left(\frac{\partial^2 w}{\partial x^2} \right) \qquad (8.13a)$$

$$M = B_{11} \left(\frac{\partial u}{\partial x} \right) - D_{11} \left(\frac{\partial^2 w}{\partial x^2} \right) \qquad (8.13b)$$

$\left(A_{11}, B_{11}, D_{11} \right) = \int\limits_A Q_{11} \left(1, z, z^2 \right) dA$ are the stiffness coefficients of FG beam.

The strain energy, kinetic energy, and work done by elastic foundation for the FG beam can be depicted as:

$$S = \frac{1}{2} \int\limits_0^L \left[A_{11} \left(\frac{\partial u}{\partial x} \right)^2 - 2B_{11} \left(\frac{\partial u}{\partial x} \right) \left(\frac{\partial^2 w}{\partial x^2} \right) + D_{11} \left(\frac{\partial^2 w}{\partial x^2} \right)^2 \right] dx \qquad (8.14)$$

$$T = \frac{1}{2} \int\limits_0^L \left[I_0 \left(\left(\frac{\partial u}{\partial t} \right)^2 + \left(\frac{\partial w}{\partial t} \right)^2 \right) - 2I_1 \left(\frac{\partial u}{\partial t} \right) \left(\frac{\partial^2 w}{\partial x \partial t} \right) + I_2 \left(\frac{\partial^2 w}{\partial x \partial t} \right)^2 \right] dx \qquad (8.15)$$

$$W = -\frac{1}{2} \int\limits_0^L \left[k_w w^2 + k_g \left(\frac{\partial w}{\partial x} \right)^2 \right] dx \qquad (8.16)$$

Assuming the motion of the FG beam as sinusoidal, i.e., plugging $u(x,t) = U(x)\cos(\omega t)$ and $w(x,t) = W(x)\cos(\omega t)$, the maximum strain energy (S_{max}), kinetic energy (T_{max}), and work done by elastic foundation (W_{max}) for the FG beam can be obtained as:

$$S_{max} = \frac{1}{2} \int\limits_0^L \left[A_{11} \left(\frac{dU}{dx} \right)^2 - 2B_{11} \left(\frac{dU}{dx} \right) \left(\frac{d^2 W}{dx^2} \right) + D_{11} \left(\frac{d^2 W}{dx^2} \right)^2 \right] dx \qquad (8.17)$$

$$T_{max} = \frac{\omega^2}{2} \int\limits_0^L \left[I_0 \left(U^2 + W^2 \right) - 2I_1 (U) \left(\frac{dW}{dx} \right) + I_2 \left(\frac{dW}{dx} \right)^2 \right] dx \qquad (8.18)$$

$$W_{max} = -\frac{1}{2} \int\limits_0^L \left[k_w W^2 + k_g \left(\frac{dW}{dx} \right)^2 \right] dx \qquad (8.19)$$

8.2.1 Modeling with Geometrical Uncertainties

Since the uncertainties are associated with geometrical properties or dimension of the specimen, incorporating the concept of the double parametric as given in preliminaries, Symmetric Gaussian Fuzzy Number (SGFN) $\tilde{L} = sgfn\left(\overline{L}, \sigma_1, \sigma_1\right)$ and $\tilde{h} = sgfn\left(\overline{h}, \sigma_2, \sigma_2\right)$ can be expressed in double parametric form as:

$$\tilde{L}(\alpha_1, \beta_1) = sgfn\left(\overline{L}, \sigma_1, \sigma_1\right) = \left(\left(2\beta_1 \sqrt{-2\sigma_1^2 \ln(\alpha_1)}\right) + \overline{L} - \sqrt{-2\sigma_1^2 \ln(\alpha_1)}\right), \qquad (8.20)$$

where $\alpha_1 \in \begin{pmatrix} 0 & 1 \end{pmatrix}$ and $\beta_1 \in \begin{bmatrix} 0 & 1 \end{bmatrix}$.

$$\tilde{h}(\alpha_2, \beta_2) = sgfn\left(\overline{h}, \sigma_2, \sigma_2\right) = \left(\left(2\beta_2 \sqrt{-2\sigma_2^2 \ln(\alpha_2)}\right) + \overline{h} - \sqrt{-2\sigma_2^2 \ln(\alpha_2)}\right), \quad (8.21)$$

where $\alpha_2 \in \begin{pmatrix} 0 & 1 \end{pmatrix}$ and $\beta_2 \in \begin{bmatrix} 0 & 1 \end{bmatrix}$.

Substituting Eq. (8.17), Eq. (8.18), and Eq. (8.19) into Lagrangian energy function $\Pi = S_{max} - W_{max} - T_{max}$ by considering geometrical uncertainties and setting $\Pi = 0$, one can get:

$$\int_0^{\tilde{L}(\alpha_1, \beta_1)} \left[\tilde{A}_{11} \left(\frac{dU}{dx}\right)^2 - 2\tilde{B}_{11} \left(\frac{d^2W}{dx^2}\right)\left(\frac{dU}{dx}\right) + \tilde{D}_{11} \left(\frac{d^2W}{dx^2}\right)^2 + k_w W^2 + k_g \left(\frac{dW}{dx}\right)^2 \right] dx$$

$$= \omega^2 \int_0^{\tilde{L}(\alpha_1, \beta_1)} \left[\tilde{I}_0 \left(U^2 + W^2\right) - 2\tilde{I}_1 (U)\left(\frac{dW}{dx}\right) + \tilde{I}_2 \left(\frac{dW}{dx}\right)^2 \right] dx$$

$$(8.22)$$

Equation (8.22) represents the energy form of equation of motion for the FG beam with geometrical uncertainties.

Likewise, substituting Eq. (8.13) into Eq. (8.12) by considering geometrical uncertainties, the governing equations of motion in terms of displacement for Navier's technique can be obtained as:

$$\tilde{A}_{11} \left(\frac{\partial^2 u}{\partial x^2}\right) - \tilde{B}_{11} \left(\frac{\partial^3 w}{\partial x^3}\right) = \tilde{I}_0 \left(\frac{\partial^2 u}{\partial t^2}\right) - \tilde{I}_1 \left(\frac{\partial^3 w}{\partial x \partial t^2}\right) \qquad (8.23a)$$

$$\tilde{B}_{11} \left(\frac{\partial^3 u}{\partial x^3}\right) - \tilde{D}_{11} \left(\frac{\partial^4 w}{\partial x^4}\right) = \tilde{I}_0 \left(\frac{\partial^2 w}{\partial t^2}\right) + \tilde{I}_1 \left(\frac{\partial^3 u}{\partial x \partial t^2}\right) - \tilde{I}_2 \left(\frac{\partial^4 w}{\partial x^2 \partial t^2}\right)$$

$$+ k_w w - k_g \left(\frac{\partial^2 w}{\partial x^2}\right)$$

$$(8.23b)$$

Where $\tilde{A}_{11}$, $\tilde{B}_{11}$, and $\tilde{D}_{11}$ are uncertain stiffness coefficients, and $\tilde{I}_0$, $\tilde{I}_1$, and $\tilde{I}_2$ are uncertain mass moments of inertia of FG beam, which are expressed as:

$$
\begin{aligned}
\tilde{A}_{11} &= \left[\frac{b \times \tilde{h}(\alpha_2, \beta_2)}{1-v^2}\right]\left[\frac{(E_c - E_m)}{k+1} + E_m\right] \\[2mm]
&= \left[\frac{b \times sgfn(\bar{h}, \sigma_2, \sigma_2)}{1-v^2}\right]\left[\frac{(E_c - E_m)}{k+1} + E_m\right] \\[2mm]
&= \left[\frac{b \times \left(\left(2\beta_2\sqrt{-2\sigma_2^2 \ln(\alpha_2)}\right) + \bar{h} - \sqrt{-2\sigma_2^2 \ln(\alpha_2)}\right)}{1-v^2}\right]\left[\frac{(E_c - E_m)}{k+1} + E_m\right]
\end{aligned}
\tag{8.24}
$$

$$
\begin{aligned}
\tilde{B}_{11} &= \frac{b \times \tilde{h}^2(\alpha_2, \beta_2) \times k}{1-v^2}\left[\frac{(E_c - E_m)}{2(k+1)(k+2)}\right] \\[2mm]
&= \frac{b \times \left(sgfn(\bar{h}, \sigma_2, \sigma_2)\right)^2 \times k}{1-v^2}\left[\frac{(E_c - E_m)}{2(k+1)(k+2)}\right] \\[2mm]
&= \left[\frac{b \times \left(\left(2\beta_2\sqrt{-2\sigma_2^2 \ln(\alpha_2)}\right) + \bar{h} - \sqrt{-2\sigma_2^2 \ln(\alpha_2)}\right)^2 \times k}{1-v^2}\right]\left[\frac{(E_c - E_m)}{2(k+1)(k+2)}\right]
\end{aligned}
\tag{8.25}
$$

$$
\begin{aligned}
\tilde{D}_{11} &= \left[\frac{b \times \tilde{h}^3(\alpha_2, \beta_2)}{1-v^2}\right]\left[\frac{(E_c - E_m)(k^2 + k + 2)}{4(k+1)(k+2)(k+3)} + \frac{E_m}{12}\right] \\[2mm]
&= \left[\frac{b \times \left(sgfn(\bar{h}, \sigma_2, \sigma_2)\right)^3}{1-v^2}\right]\left[\frac{(E_c - E_m)(k^2 + k + 2)}{4(k+1)(k+2)(k+3)} + \frac{E_m}{12}\right] \\[2mm]
&= \left[\frac{b \times \left(\left(2\beta_2\sqrt{-2\sigma_2^2 \ln(\alpha_2)}\right) + \bar{h} - \sqrt{-2\sigma_2^2 \ln(\alpha_2)}\right)^3}{1-v^2}\right] \\[2mm]
&\quad \left[\frac{(E_c - E_m)(k^2 + k + 2)}{4(k+1)(k+2)(k+3)} + \frac{E_m}{12}\right]
\end{aligned}
\tag{8.26}
$$

$$\tilde{I}_0 = \left[b \times \tilde{h}(\alpha_2, \beta_2) \right] \left[\frac{(\rho_c - \rho_m)}{k+1} + \rho_m \right]$$

$$= \left[b \times sgfn\left(\overline{h}, \sigma_2, \sigma_2 \right) \right] \left[\frac{(\rho_c - \rho_m)}{k+1} + \rho_m \right] \qquad (8.27)$$

$$= \left[b \times \left(\left(2\beta_2 \sqrt{-2\sigma_2^2 \ln(\alpha_2)} \right) + \overline{h} - \sqrt{-2\sigma_2^2 \ln(\alpha_2)} \right) \right] \left[\frac{(\rho_c - \rho_m)}{k+1} + \rho_m \right]$$

$$\tilde{I}_1 = \left[b \times \tilde{h}^2(\alpha_2, \beta_2) \times k \right] \left[\frac{(\rho_c - \rho_m)}{2(k+1)(k+2)} \right]$$

$$= \left[b \times \left(sgfn\left(\overline{h}, \sigma_2, \sigma_2 \right) \right)^2 \times k \right] \left[\frac{(\rho_c - \rho_m)}{2(k+1)(k+2)} \right] \qquad (8.28)$$

$$= \left[b \times \left(\left(\left(2\beta_2 \sqrt{-2\sigma_2^2 \ln(\alpha_2)} \right) + \overline{h} - \sqrt{-2\sigma_2^2 \ln(\alpha_2)} \right) \right)^2 \times k \right] \left[\frac{(\rho_c - \rho_m)}{2(k+1)(k+2)} \right]$$

$$\tilde{I}_2 = \left[b \times \tilde{h}^3(\alpha_2, \beta_2) \right] \left[\frac{(\rho_c - \rho_m)(k^2 + k + 2)}{4(k+1)(k+2)(k+3)} + \frac{\rho_m}{12} \right]$$

$$= \left[b \times \left(sgfn\left(\overline{h}, \sigma_2, \sigma_2 \right) \right)^3 \right] \left[\frac{(\rho_c - \rho_m)(k^2 + k + 2)}{4(k+1)(k+2)(k+3)} + \frac{\rho_m}{12} \right] \qquad (8.29)$$

$$= \left[b \times \left(\left(\left(2\beta_2 \sqrt{-2\sigma_2^2 \ln(\alpha_2)} \right) + \overline{h} - \sqrt{-2\sigma_2^2 \ln(\alpha_2)} \right) \right)^3 \right]$$

$$\left[\frac{(\rho_c - \rho_m)(k^2 + k + 2)}{4(k+1)(k+2)(k+3)} + \frac{\rho_m}{12} \right]$$

8.3 SOLUTION PROCEDURES

In the upcoming subsections, Navier's technique and Hermite–Ritz method have been described to solve the governing equations of motion for the proposed model. Navier's method has been used for hinged–hinged boundary condition, whereas Hermite–Ritz method is being employed for hinged–hinged, clamped–hinged, and clamped–clamped boundary conditions.

8.3.1 Application of Navier's Technique

As per Navier's technique, the axial displacement $u(x,t)$ and transverse displacement $w(x,t)$ can be expanded in terms of sine and cosine series as [1]:

$$u(x,t) = \sum_{m=1}^{\infty} u_m \cos\left(\frac{m\pi}{\tilde{L}}x\right) e^{i\omega t} \tag{8.30a}$$

$$w(x,t) = \sum_{m=1}^{\infty} w_m \sin\left(\frac{m\pi}{\tilde{L}}x\right) e^{i\omega t} \tag{8.30b}$$

Where u_m and w_m are arbitrary parameters, and ω is the natural frequency of vibration. Plugging Eq. (8.30) into Eq. (8.23), generalized eigenvalue problem for the free vibration of FG beam with material uncertainties will be obtained as:

$$[K]\{X\} = \omega^2 [M]\{X\} \tag{8.31}$$

Here, $[K] = \begin{bmatrix} k_{11} & k_{12} \\ k_{21} & k_{22} \end{bmatrix}$, $[M] = \begin{bmatrix} m_{11} & m_{12} \\ m_{21} & m_{22} \end{bmatrix}$, and $\{X\} = \begin{bmatrix} u_m & w_m \end{bmatrix}^T$,

Where,

$$k_{11} = -\tilde{A}_{11}\left(\frac{m\pi}{\tilde{L}}\right)^2, k_{12} = k_{21} = \tilde{B}_{11}\left(\frac{m\pi}{\tilde{L}}\right)^3, k_{22} = -\tilde{D}_{11}\left(\frac{m\pi}{\tilde{L}}\right)^4 - (k_w) - (k_g)\left(\frac{m\pi}{\tilde{L}}\right)^2,$$

$$m_{11} = -\tilde{I}_0, m_{12} = m_{21} = \tilde{I}_1\left(\frac{m\pi}{\tilde{L}}\right), m_{22} = -\tilde{I}_0 - \tilde{I}_2\left(\frac{m\pi}{\tilde{L}}\right)^2$$

By solving Eq. (8.31), the natural frequencies for the proposed model will be determined for uncertain parameters of the hinged–hinged (HH) boundary condition. It is worth noting that by substituting $\alpha = 1$ in Eq. (8.31), this uncertain model can be converted into a deterministic one.

8.3.2 Application of Hermite–Ritz Method

The axial displacement $U(X)$ and transverse displacement $W(X)$ can be now expressed as [1]:

$$U(X) = X^\eta (L - X)^\kappa \sum_{i=1}^{n} c_i H(i-1, X) \tag{8.32a}$$

$$W(X) = X^\eta (L - X)^\kappa \sum_{i=1}^{n} d_i H(i-1, X) \tag{8.32b}$$

Here $c_i's$ and $d_i's$ are unknown coefficients, $H(n, X)$ is the nth term of Hermite polynomial which is used shape function, and $X^\eta (R - X)^\kappa$ is the admissible function with exponents η and κ. For hinged–hinged (HH), clamped–hinged (CH), and clamped–clamped (CC) boundary conditions, we have $(\eta = \kappa = 1)$, $(\eta = 2, \kappa = 1)$, and $(\eta = \kappa = 2)$, respectively.

Substituting Eq. (8.32) into Eq. (8.22) and differentiating partially with respect to the unknown coefficients $c_i's$ and $d_i's$, $i = 1,2,3...n$, yield generalized eigenvalue problem as:

$$[K]\{X\} = \omega^2 [M]\{X\} \qquad (8.33)$$

where $\{X\} = [c_1,c_2,c_3,...c_n,d_1,d_2,d_3,...d_n]^T$, $[K]$ represents the stiffness matrix, and $[M]$ denotes the mass matrix.

8.4   NUMERICAL RESULTS AND DISCUSSION

The FG beam is assumed in this study to be made up of metal constituent such as alpha-beta titanium alloy (Ti-6AL-4V) and ceramic constituent such as zirconium dioxide (ZrO_2). The mechanical characteristics of the FG beam are listed in Table 8.1a, and the schematic representation is shown in Figure 8.1a.

TABLE 8.1a

Material Properties [6] of Alpha-Beta Titanium Alloy (Ti-6AL-4V) and Zirconium Dioxide (ZrO_2)

Properties	Ti-6AL-4V (Metal)	ZrO2 (Ceramic)
Young's modulus	$E_L = 105.7\,GPa$	$E_U = 151\,GPa$
Material density	$\rho_L = 4429\,kg.m^{-3}$	$\rho_U = 3000\,kg.m^{-3}$

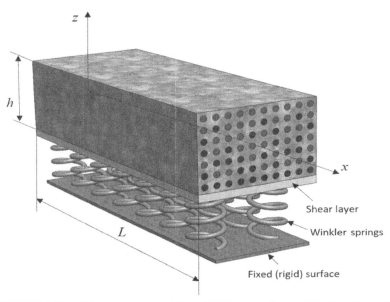

FIGURE 8.1A   Schematic representation of FG beam resting on Winkler–Pasternak elastic foundation, where upper layer represents purely Ceramic (ZrO_2) and lower layer represents purely Metal (Ti-6AL-4V).

The geometrical properties of FG beam are taken as uncertain in term of SGFN which is given in Table 8.1b, and the geometrical representation is depicted in Figure 8.1b. It's worth noting that the modal value and standard deviation for length correspond to 8,000 mm and a worst-case of ±30%, whereas for the thickness, the modal value is 100 nm with a worst-case of standard deviation ±9% and the breadth of the FG beam is taken as 400 nm. Different intervals are constructed for different values of α, and the lower bound (LB) and upper bound (UB) of first four modes of natural frequencies are found for each interval by incorporating the double parametric form-based Navier's technique and Hermite–Ritz method for HH, CH, and CC boundary conditions. Table 8.2 shows the intervals that are considered in this chapter for the investigation. Also, the study is conducted in three ways, i.e., the results are generated by considering: (i) The length (L) as SGFN, (ii) the thickness (h) as SGFN, and, finally, (iii) L and h as SGFNs.

TABLE 8.1b

Parameter Details in Terms of Symmetric Gaussian Fuzzy Number (SGFN)

Parameters	Modal Value	Sym. Gaussian fuzzy No.
Length (L)	$\bar{L} = 8,000\,\text{mm}$	$\tilde{L} = sgfn\left(\bar{L}, 10\%\bar{L}, 10\%\bar{L}\right)$
Thickness (h)	$\bar{h} = 100\,\text{mm}$	$\tilde{h} = sgfn\left(\bar{h}, 3\%\bar{h}, 3\%\bar{h}\right)$

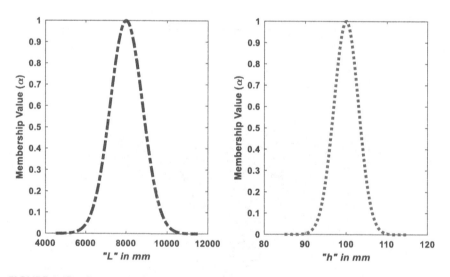

FIGURE 8.1B   Symmetric Gaussian Fuzzy Numbers (SGFNs) associated with "length (L)", and "thickness (h)".

TABLE 8.2
Uncertain Parameters for Different α Values

α	$\tilde{L} = \left[\underline{L}(\alpha), \overline{L}(\alpha)\right]$	$\tilde{h} = \left[\underline{h}(\alpha), \overline{h}(\alpha)\right]$
0.1	[6283.2 9716.8]	[93.562 106.43]
0.2	[6564.7 9435.3]	[94.617 105.38]
0.3	[6758.6 9241.4]	[95.344 104.65]
0.4	[6917.0 9083.0]	[95.938 104.06]
0.5	[7058.1 8941.9]	[96.467 103.53]
0.6	[7191.4 8808.6]	[96.967 103.03]
0.7	[7324.3 8675.7]	[97.466 102.53]
0.8	[7465.6 8534.4]	[97.995 102.00]
0.9	[7632.8 8.3672]	[98.622 101.37]
1	$\overline{L} = 8000$	$\overline{h} = 100$

8.4.1 VALIDATION OF PROPOSED MODEL THROUGH DOUBLE PARAMETRIC FORM-BASED NAVIER'S METHOD AND HERMITE–RITZ METHOD

The present model is validated by using double parametric form-based Navier's method and Hermite–Ritz method for HH boundary condition. The lower bound (LB) and upper bound (UB) for first four modes of natural frequencies are taken into account by considering three cases of (i) the length (L) as uncertain or fuzzy, (ii) the thickness (h) as uncertain or fuzzy, and, finally, (iii) both the length (L) and thickness (h) as uncertain or fuzzy. Tabular results for each of the above cases are demonstrated in Tables 8.3–8.8. Here, calculations are done by taking power-law exponent $k = 1$ and nondimensional Winkler and Pasternak elastic parameters as $K_w = K_p = 40$. Tables 8.3–8.5 illustrate the LB and UB of the natural frequencies (Hz) by considering the above-mentioned three cases, respectively, by incorporating double parametric form-based Navier's method. Likewise, Tables 8.6–8.8 elucidate the natural frequencies by utilizing double parametric form-based Hermite–Ritz method. Based on these results, it is clear that the natural frequencies of the present uncertain model are well validated. It is also worth mentioning that Navier's approach takes substantially less time than the Hermite–Ritz method.

TABLE 8.3

Natural Frequencies in Hz for HH Boundary Condition with
$\tilde{L} = sgfn\left(\overline{L}, 10\%\overline{L}, 10\%\overline{L}\right)$

	First Mode		Second Mode		Third mode		Fourth Mode	
α	UB	LB	UB	LB	UB	LB	UB	LB
0.1	14.7339	6.1612	37.6292	15.7382	73.4149	30.7150	122.8357	51.4139
0.2	13.4977	6.5343	34.4728	16.6911	67.2597	32.5741	112.5441	54.5248
0.3	12.7344	6.8114	32.5240	17.3986	63.4591	33.9547	106.1888	56.8349
0.4	12.1578	7.0510	31.0517	18.0107	60.5879	35.1489	101.3872	58.8332
0.5	11.6768	7.2752	29.8234	18.5833	58.1922	36.2660	97.3807	60.7024
0.6	11.2479	7.4971	28.7283	19.1499	56.0562	37.3715	93.8084	62.5520
0.7	10.8433	7.7286	27.6953	19.7412	54.0413	38.5250	90.4383	64.4819
0.8	10.4369	7.9865	26.6576	20.3998	52.0172	39.8100	87.0528	66.6319
0.9	9.9847	8.3089	25.5028	21.2231	49.7647	41.4161	83.2850	69.3189
1	9.0891	9.0891	23.2157	23.2157	45.3034	45.3034	75.8221	75.8221

TABLE 8.4

Natural Frequencies in Hz for HH Boundary Condition with
$\tilde{h} = sgfn\left(\overline{h}, 3\%\overline{h}, 3\%\overline{h}\right)$

	First Mode		Second Mode		Third mode		Fourth Mode	
α	LB	UB	LB	UB	LB	UB	LB	UB
0.1	8.5041	9.6742	21.7219	24.7095	42.3900	48.2162	70.9502	80.6921
0.2	8.6000	9.5783	21.9668	24.4646	42.8677	47.7386	71.7491	79.8938
0.3	8.6661	9.5122	22.1355	24.2959	43.1967	47.4097	72.2994	79.3439
0.4	8.7201	9.4582	22.2734	24.1580	43.4656	47.1409	72.7490	78.8945
0.5	8.7681	9.4101	22.3961	24.0353	43.7050	46.9016	73.1493	78.4944
0.6	8.8136	9.3647	22.5121	23.9193	43.9312	46.6754	73.5276	78.1162
0.7	8.8589	9.3194	22.6278	23.8037	44.1568	46.4499	73.9049	77.7391
0.8	8.9070	9.2713	22.7507	23.6808	44.3965	46.2102	74.3057	77.3384
0.9	8.9640	9.2143	22.8962	23.5353	44.6802	45.9265	74.7801	76.8641
1	9.0891	9.0891	23.2157	23.2157	45.3034	45.3034	75.8221	75.8221

TABLE 8.5

Natural Frequencies in Hz for HH Boundary Condition with
$\tilde{L} = sgfn\left(\overline{L}, 10\%\overline{L}, 10\%\overline{L}\right)$, and $\tilde{h} = sgfn\left(\overline{h}, 3\%\overline{h}, 3\%\overline{h}\right)$

α	First Mode		Second Mode		Third mode		Fourth Mode	
	UB	LB	UB	LB	UB	LB	UB	LB
0.1	13.7856	6.5578	35.2086	16.7510	68.6968	32.6906	114.9523	54.7186
0.2	12.7713	6.8860	32.6187	17.5891	63.6455	34.3257	106.5041	57.4547
0.3	12.1417	7.1284	31.0109	18.2083	60.5095	35.5339	101.2591	59.4763
0.4	11.6641	7.3373	29.7915	18.7419	58.1310	36.5750	97.2806	61.2183
0.5	11.2644	7.5322	28.7707	19.2394	56.1397	37.5458	93.9499	62.8427
0.6	10.9069	7.7244	27.8577	19.7304	54.3588	38.5036	90.9709	64.4453
0.7	10.5686	7.9244	26.9940	20.2411	52.6739	39.5001	88.1524	66.1127
0.8	10.2278	8.1465	26.1236	20.8085	50.9761	40.6070	85.3123	67.9647
0.9	9.8472	8.4233	25.1518	21.5152	49.0803	41.9858	82.1407	70.2716
1	9.0891	9.0891	23.2157	23.2157	45.3034	45.3034	75.8221	75.8221

TABLE 8.6

Natural Frequencies in Hz for HH Boundary Condition with
$\tilde{L} = sgfn\left(\overline{L}, 10\%\overline{L}, 10\%\overline{L}\right)$

α	First Mode		Second Mode		Third mode		Fourth Mode	
	UB	LB	UB	LB	UB	LB	UB	LB
0.1	14.7483	6.1672	37.6289	15.7382	73.4405	30.7257	122.8329	51.4138
0.2	13.5108	6.5406	34.4725	16.6910	67.2832	32.5855	112.5419	54.5247
0.3	12.7467	6.8180	32.5237	17.3986	63.4813	33.9666	106.1868	56.8348
0.4	12.1696	7.0579	31.0515	18.0107	60.6090	35.1612	101.3855	58.8330
0.5	11.6881	7.2823	29.8232	18.5832	58.2125	36.2787	97.3792	60.7021
0.6	11.2588	7.5044	28.7281	19.1499	56.0758	37.3846	93.8070	62.5518
0.7	10.8538	7.7361	27.6951	19.7411	54.0602	38.5385	90.4372	64.4816
0.8	10.4471	7.9942	26.6574	20.3997	52.0354	39.8240	87.0517	66.6315
0.9	9.9944	8.3169	25.5026	21.2230	49.7821	41.4306	83.2841	69.3184
1	9.0980	9.0980	23.2156	23.2156	45.3193	45.3193	75.8215	75.8215

TABLE 8.7

Natural Frequencies in Hz for HH Boundary Condition with
$\tilde{h} = sgfn\left(\bar{h}, 3\%\bar{h}, 3\%\bar{h}\right)$

	First Mode		Second Mode		Third mode		Fourth Mode	
α	LB	UB	LB	UB	LB	UB	LB	UB
0.1	8.5123	9.6836	21.7218	24.7093	42.4048	48.2330	70.9498	80.6912
0.2	8.6083	9.5876	21.9667	24.4644	42.8827	47.7553	71.7486	79.8929
0.3	8.6745	9.5214	22.1354	24.2957	43.2119	47.4263	72.2989	79.3430
0.4	8.7285	9.4674	22.2733	24.1579	43.4808	47.1574	72.7485	78.8937
0.5	8.7766	9.4193	22.3960	24.0352	43.7203	46.9180	73.1488	78.4936
0.6	8.8221	9.3738	22.5120	23.9192	43.9466	46.6918	73.5271	78.1154
0.7	8.8675	9.3284	22.6277	23.8035	44.1723	46.4661	73.9043	77.7383
0.8	8.9156	9.2803	22.7506	23.6806	44.4120	46.2264	74.3051	77.3377
0.9	8.9727	9.2232	22.8961	23.5351	44.6959	45.9426	74.7795	76.8633
1	9.0980	9.0980	23.2156	23.2156	45.3193	45.3193	75.8215	75.8215

TABLE 8.8

Natural Frequencies in Hz for HH Boundary Condition with
$\tilde{L} = sgfn\left(\bar{L}, 10\%\bar{L}, 10\%\bar{L}\right)$, **and** $\tilde{h} = sgfn\left(\bar{h}, 3\%\bar{h}, 3\%\bar{h}\right)$

	First Mode		Second Mode		Third mode		Fourth Mode	
α	UB	LB	UB	LB	UB	LB	UB	LB
0.1	13.7990	6.5642	35.2083	16.7510	68.7208	32.7021	114.9502	54.7184
0.2	12.7837	6.8926	32.6185	17.5890	63.6677	34.3378	106.5024	57.4545
0.3	12.1535	7.1353	31.0107	18.2082	60.5307	35.5463	101.2575	59.4760
0.4	11.6755	7.3445	29.7913	18.7418	58.1513	36.5878	97.2792	61.2180
0.5	11.2753	7.5395	28.7705	19.2393	56.1593	37.5589	93.9486	62.8423
0.6	10.9174	7.7319	27.8575	19.7303	54.3779	38.5171	90.9698	64.4450
0.7	10.5789	7.9321	26.9938	20.2410	52.6923	39.5139	88.1514	66.1123
0.8	10.2377	8.1544	26.1235	20.8084	50.9940	40.6212	85.3113	67.9643
0.9	9.8568	8.4314	25.1516	21.5151	49.0974	42.0005	82.1398	70.2711
1	9.0980	9.0980	23.2156	23.2156	45.3193	45.3193	75.8215	75.8215

8.4.2 Effects of Geometrical Uncertainties on Natural Frequencies

Material uncertainties in FG structures are caused by atomic structure defects and manufacturing abnormalities, which have a significant impact on the dynamical characteristics. In this subsection, we'll look at the impact of geometrical uncertainty on the natural frequencies of FG beam. As shown in Table 8.1b and Figure 8.1b, the uncertain input variables are treated as Symmetric Gaussian Fuzzy

Numbers (SGFNs). The first four natural frequencies of the FG beams are calculated using the double parametric form-based Navier's technique and Hermite–Ritz method for HH, CH, and CC boundary conditions, as shown in Tables 8.1–8.14 as tabular results for all three cases: (i) The length (L) as uncertain or fuzzy, (ii) the thickness (h) as uncertain or fuzzy, and, finally, (iii) both the length (L) and thickness (h) as uncertain or fuzzy. Figures 8.2–8.13 illustrate the Gaussian fuzzy output for fundamental natural frequencies for all the undertaken boundary conditions. In addition, a comparison study has been done using the Gaussian fuzzy number to represent the fundamental natural frequency, and the findings were shown as graphical results in Figures 8.5, 8.9, and 8.13. When the length is uncertain, the spreads or fuzziness of Gaussian fuzzy output of natural frequency is the largest, but when the thickness is uncertain, the fuzziness of the frequency is the lowest. In addition, for "uncertain length" and "uncertain length and thickness" as input, the fuzzy output of natural frequency follows the same pattern, whereas it is bit different in case of thickness as uncertain input. Furthermore, in the case of uncertain thickness, the fuzzy output of natural frequency appears symmetrical, which is not the case in the other two cases.

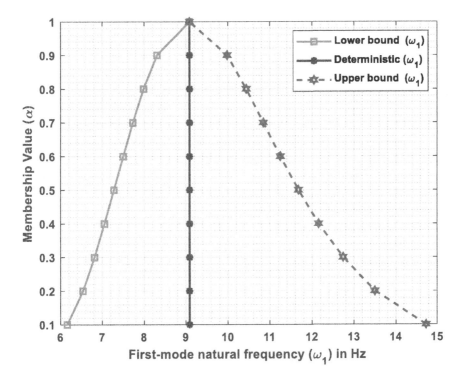

FIGURE 8.2 Gaussian fuzzy output for fundamental natural frequency with $\tilde{L} = sgfn\left(\overline{L}, 10\%\,\overline{L}, 10\%\,\overline{L}\right)$ for HH boundary condition.

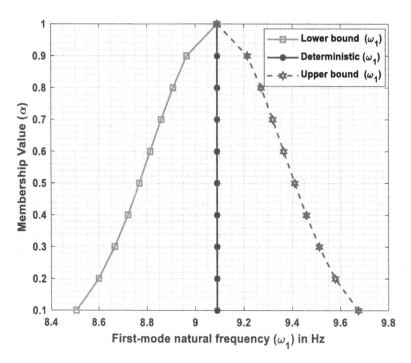

FIGURE 8.3 Gaussian fuzzy output for fundamental natural frequency with $\tilde{h} = sgfn\left(\overline{h}, 3\%\overline{h}, 3\%\overline{h}\right)$ for HH boundary condition.

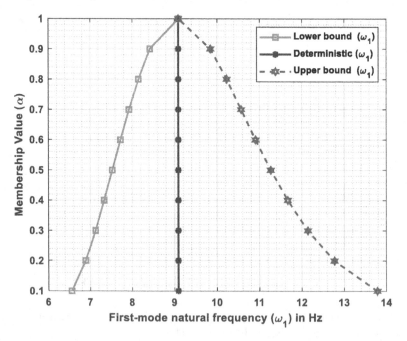

FIGURE 8.4 Gaussian fuzzy output for fundamental natural frequency with $\tilde{L} = sgfn\left(\overline{L}, 10\%\overline{L}, 10\%\overline{L}\right)$ and $\tilde{h} = sgfn\left(\overline{h}, 3\%\overline{h}, 3\%\overline{h}\right)$ for HH boundary condition.

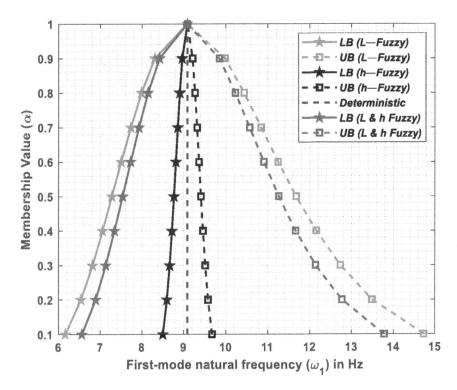

FIGURE 8.5 Comparisons of Gaussian fuzzy output for fundamental natural frequency for HH boundary condition.

TABLE 8.9

Natural Frequencies in Hz for CC Boundary Condition with
$\tilde{L} = sgfn\left(\bar{L}, 10\%\bar{L}, 10\%\bar{L}\right)$

	First Mode		Second Mode		Third mode		Fourth Mode	
α	UB	LB	UB	LB	UB	LB	UB	LB
0.1	21.3732	8.9377	51.5293	21.5527	94.6666	39.6086	151.3711	63.3630
0.2	19.5799	9.4788	47.2072	22.8575	86.7304	42.0059	138.6904	67.1968
0.3	18.4727	9.8807	44.5386	23.8265	81.8300	43.7862	130.8596	70.0436
0.4	17.6363	10.2284	42.5226	24.6647	78.1279	45.3261	124.9431	72.5060
0.5	16.9385	10.5536	40.8406	25.4488	75.0389	46.7666	120.0064	74.8094
0.6	16.3163	10.8755	39.3410	26.2247	72.2848	48.1921	115.6046	77.0888
0.7	15.7295	11.2113	37.9264	27.0343	69.6868	49.6794	111.4520	79.4670
0.8	15.1400	11.5854	36.5054	27.9363	67.0769	51.3365	107.2802	82.1164
0.9	14.4840	12.0530	34.9241	29.0636	64.1725	53.4074	102.6374	85.4275
1	13.1849	13.1849	31.7923	31.7923	58.4199	58.4199	93.4413	93.4413

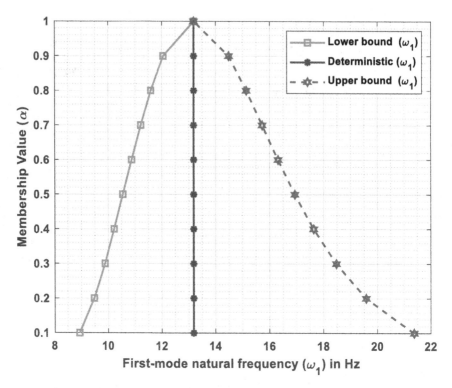

FIGURE 8.6 Gaussian fuzzy output for fundamental natural frequency with $\tilde{L} = sgfn\left(\bar{L}, 10\%\,\bar{L}, 10\%\,\bar{L}\right)$ for CC boundary condition.

TABLE 8.10

Natural Frequencies in Hz for CC Boundary Condition with
$\tilde{h} = sgfn\left(\bar{h}, 3\%\,\bar{h}, 3\%\,\bar{h}\right)$

	First Mode		Second Mode		Third mode		Fourth Mode	
α	LB	UB	LB	UB	LB	UB	LB	UB
0.1	12.3362	14.0336	29.7467	33.8377	54.6634	62.1754	87.4382	99.4418
0.2	12.4753	13.8944	30.0821	33.5024	55.2794	61.5598	88.4226	98.4581
0.3	12.5712	13.7986	30.3132	33.2714	55.7037	61.1356	89.1007	97.7805
0.4	12.6495	13.7203	30.5019	33.0826	56.0503	60.7891	89.6547	97.2269
0.5	12.7192	13.6505	30.6700	32.9146	56.3590	60.4805	90.1480	96.7339
0.6	12.7851	13.5846	30.8289	32.7558	56.6507	60.1889	90.6141	96.2679
0.7	12.8509	13.5189	30.9872	32.5974	56.9415	59.8981	91.0789	95.8033
0.8	12.9207	13.4491	31.1555	32.4291	57.2506	59.5891	91.5728	95.3096
0.9	13.0033	13.3664	31.3548	32.2299	57.6164	59.2233	92.1574	94.7251
1	13.1849	13.1849	31.7923	31.7923	58.4199	58.4199	93.4413	93.4413

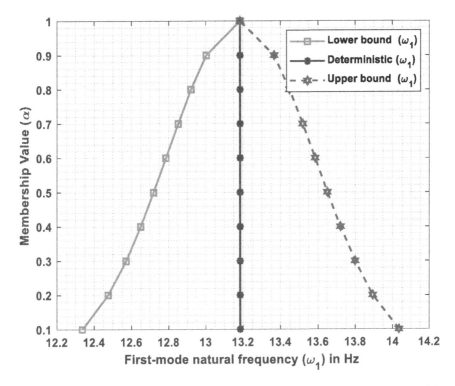

FIGURE 8.7 Gaussian fuzzy output for fundamental natural frequency with $\tilde{h} = sgfn\left(\overline{h}, 3\%\overline{h}, 3\%\overline{h}\right)$ for CC boundary condition.

TABLE 8.11

Natural Frequencies in Hz for CC Boundary Condition with
$\tilde{L} = sgfn\left(\overline{L}, 10\%\overline{L}, 10\%\overline{L}\right)$ and $\tilde{h} = sgfn\left(\overline{h}, 3\%\overline{h}, 3\%\overline{h}\right)$

	First Mode		Second Mode		Third mode		Fourth Mode	
α	UB	LB	UB	LB	UB	LB	UB	LB
0.1	19.9975	9.5130	48.2149	22.9396	88.5839	42.1560	141.6589	67.4353
0.2	18.5262	9.9890	44.6684	24.0872	82.0707	44.2645	131.2491	70.8070
0.3	17.6129	10.3406	42.4668	24.9351	78.0272	45.8224	124.7860	73.2983
0.4	16.9202	10.6437	40.7970	25.6658	74.9602	47.1649	119.8837	75.4450
0.5	16.3403	10.9263	39.3991	26.3472	72.3927	48.4167	115.7794	77.4467
0.6	15.8217	11.2052	38.1489	27.0195	70.0964	49.6518	112.1086	79.4217
0.7	15.3310	11.4953	36.9661	27.7189	67.9238	50.9368	108.6355	81.4764
0.8	14.8366	11.8175	35.7743	28.4959	65.7346	52.3641	105.1357	83.7587
0.9	14.2846	12.2190	34.4435	29.4636	63.2900	54.1420	101.2275	86.6015
1	13.1849	13.1849	31.7923	31.7923	58.4199	58.4199	93.4413	93.4413

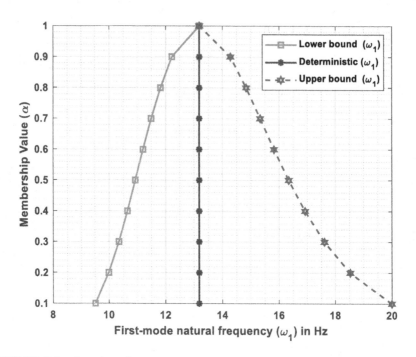

FIGURE 8.8 Gaussian fuzzy output for fundamental natural frequency with $\tilde{L} = sgfn\left(\overline{L}, 10\%\,\overline{L}, 10\%\,\overline{L}\right)$ and $\tilde{h} = sgfn\left(\overline{h}, 3\%\,\overline{h}, 3\%\,\overline{h}\right)$ for CC boundary condition.

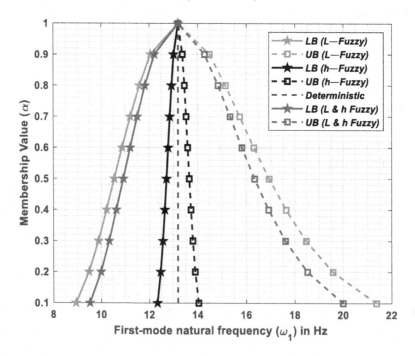

FIGURE 8.9 Comparisons of Gaussian fuzzy output for fundamental natural frequency for CC boundary condition.

TABLE 8.12

Natural Frequencies in Hz for CH Boundary Condition with
$\tilde{L} = sgfn\left(\overline{L}, 10\%\,\overline{L}, 10\%\,\overline{L}\right)$

α	First Mode UB	First Mode LB	Second Mode UB	Second Mode LB	Third mode UB	Third mode LB	Fourth Mode UB	Fourth Mode LB
0.1	17.6202	7.3682	44.0990	18.4446	83.5691	34.9645	136.6418	57.1956
0.2	16.1418	7.8144	40.4000	19.5612	76.5630	37.0808	125.1945	60.6562
0.3	15.2290	8.1457	38.1161	20.3905	72.2369	38.6523	118.1253	63.2260
0.4	14.5395	8.4323	36.3908	21.1078	68.9687	40.0117	112.7844	65.4489
0.5	13.9642	8.7004	34.9513	21.7788	66.2418	41.2834	108.3279	67.5282
0.6	13.4513	8.9658	33.6680	22.4429	63.8105	42.5418	104.3542	69.5857
0.7	12.9675	9.2426	32.4573	23.1358	61.5169	43.8547	100.6056	71.7325
0.8	12.4815	9.5511	31.2412	23.9077	59.2129	45.3175	96.8397	74.1241
0.9	11.9407	9.9366	29.8879	24.8725	56.6489	47.1457	92.6486	77.1131
1	10.8697	10.8697	27.2077	27.2077	51.5706	51.5706	84.3472	84.3472

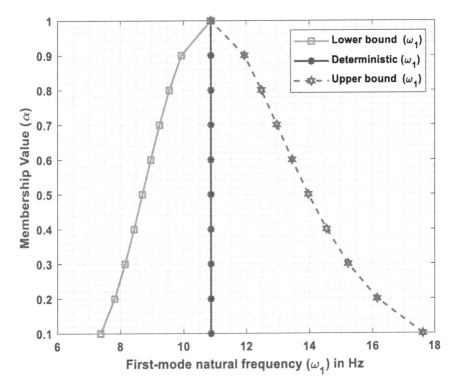

FIGURE 8.10 Gaussian fuzzy output for fundamental natural frequency with $\tilde{L} = sgfn\left(\overline{L}, 10\%\,\overline{L}, 10\%\,\overline{L}\right)$ for CH boundary condition.

TABLE 8.13

Natural Frequencies in Hz for CH Boundary Condition with
$\tilde{h} = sgfn\left(\bar{h}, 3\%\bar{h}, 3\%\bar{h}\right)$

	First Mode		Second Mode		Third mode		Fourth Mode	
α	LB	UB	LB	UB	LB	UB	LB	UB
0.1	10.1700	11.5694	25.4570	28.9582	48.2544	54.8861	78.9280	89.7640
0.2	10.2847	11.4546	25.7441	28.6712	48.7981	54.3425	79.8166	88.8761
0.3	10.3638	11.3756	25.9418	28.4735	49.1727	53.9681	80.4287	88.2643
0.4	10.4283	11.3111	26.1034	28.3120	49.4787	53.6622	80.9289	87.7645
0.5	10.4858	11.2536	26.2472	28.1682	49.7512	53.3898	81.3741	87.3195
0.6	10.5401	11.1992	26.3831	28.0322	50.0087	53.1323	81.7950	86.8988
0.7	10.5943	11.1451	26.5187	27.8967	50.2655	52.8756	82.2146	86.4794
0.8	10.6519	11.0875	26.6627	27.7527	50.5383	52.6028	82.6604	86.0337
0.9	10.7200	11.0194	26.8332	27.5822	50.8613	52.2799	83.1881	85.5061
1	10.8697	10.8697	27.2077	27.2077	51.5706	51.5706	84.3472	84.3472

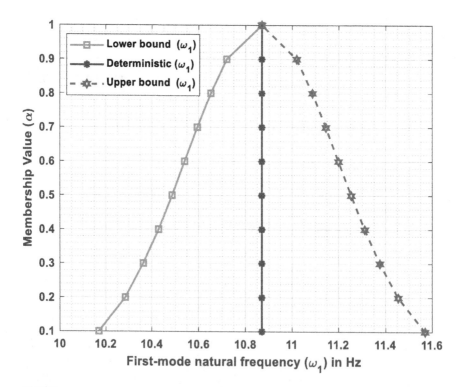

FIGURE 8.11 Gaussian fuzzy output for fundamental natural frequency with $\tilde{h} = sgfn\left(\bar{h}, 3\%\bar{h}, 3\%\bar{h}\right)$ for CH boundary condition.

TABLE 8.14

Natural Frequencies in Hz for CH Boundary Condition with
$\tilde{L} = sgfn\left(\overline{L}, 10\%\overline{L}, 10\%\overline{L}\right)$ **and** $\tilde{h} = sgfn\left(\overline{h}, 3\%\overline{h}, 3\%\overline{h}\right)$

	First Mode		Second Mode		Third mode		Fourth Mode	
α	UB	LB	UB	LB	UB	LB	UB	LB
0.1	16.4861	7.8425	41.2623	19.6315	78.1991	37.2133	127.8738	60.8717
0.2	15.2732	8.2349	38.2272	20.6136	72.4492	39.0746	118.4766	63.9153
0.3	14.5202	8.5249	36.3430	21.3393	68.8796	40.4498	112.6423	66.1641
0.4	13.9491	8.7747	34.9140	21.9646	66.1722	41.6350	108.2169	68.1020
0.5	13.4710	9.0077	33.7177	22.5477	63.9056	42.7400	104.5119	69.9089
0.6	13.0435	9.2376	32.6477	23.1231	61.8784	43.8303	101.1982	71.6917
0.7	12.6390	9.4768	31.6355	23.7217	59.9605	44.9647	98.0630	73.5465
0.8	12.2314	9.7424	30.6155	24.3866	58.0279	46.2247	94.9037	75.6066
0.9	11.7763	10.0734	29.4766	25.2148	55.8698	47.7942	91.3758	78.1728
1	10.8697	10.8697	27.2077	27.2077	51.5706	51.5706	84.3472	84.3472

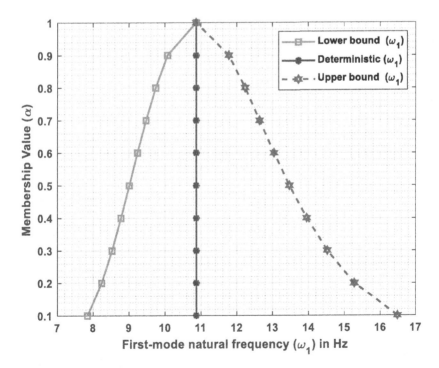

FIGURE 8.12 Gaussian fuzzy output for fundamental natural frequency with $\tilde{L} = sgfn\left(\overline{L}, 10\%\,\overline{L}, 10\%\,\overline{L}\right)$ and $\tilde{h} = sgfn\left(\overline{h}, 3\%\overline{h}, 3\%\overline{h}\right)$ for CH boundary condition.

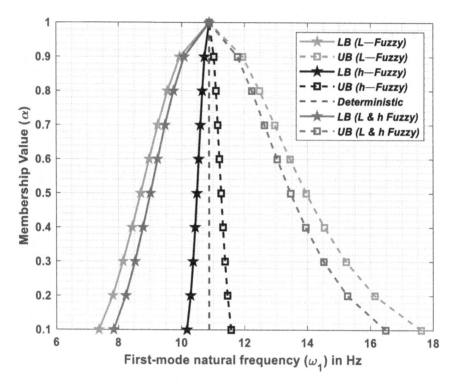

FIGURE 8.13 Comparisons of Gaussian fuzzy output for fundamental natural frequency for CH boundary condition.

8.5   CONCLUDING REMARKS

In this chapter, the effects of material uncertainties on the natural frequency of FG beam have been examined using a non-probabilistic concept, employing the double parametric form-based Navier's method and Hermite–Ritz method. The uncertainties are associated with the length and thickness of the FG beam, which are assumed in terms of Symmetric Gaussian Fuzzy Number. The lower and upper bounds of the first four natural frequencies of the uncertain systems are determined using a double parametric form-based Navier's approach and Hermite–Ritz method for HH, CH, and CC boundary conditions. Natural frequencies are obtained using Navier's method and Hermite–Ritz method in terms of the lower and upper bounds, exhibiting strong agreement, and are used to validate the uncertain model. A parametric analysis is also conducted to investigate the fuzziness or spreads of natural frequencies in relation to various uncertain parameters. It's worth mentioning that Navier's approach for HH boundary condition requires much less time than Hermite–Ritz method. Whenever the length is uncertain, the spreads or the fuzziness of the Gaussian fuzzy output of natural frequency is the highest, but the fuzziness of the frequency is the lowest when the thickness is uncertain. Furthermore, the fuzzy output of natural frequency follows the same pattern for "uncertain length" and "uncertain length and thickness", although it is a little different for thickness as an uncertain input. In addition, the fuzzy output of natural frequency seems symmetrical in the case of uncertain thickness, whereas not in the two other cases.

BIBLIOGRAPHY

[1] S.K. Jena, S. Chakraverty, M. Malikan, H. Sedighi, Implementation of Hermite—Ritz method and Navier's technique for vibration of functionally graded porous nanobeam embedded in Winkler—Pasternak elastic foundation using bi-Helmholtz nonlocal elasticity. J. Mech. Mater. Struct. **15**(3), 405–434 (2020)

[2] N Wattanasakulpong, V Ungbhakorn, Linear and nonlinear vibration analysis of elastically restrained ends FGM beams with porosities. Aerosp. Sci. Technol. **32**, 111–120 (2014)

[3] D. Shahsavari, M. Shahsavari, L. Li, B. Karami, A novel quasi-3D hyperbolic theory for free vibration of FG plates with porosities resting on Winkler/Pasternak/Kerr foundation. Aerosp. Sci. Technol. **72**, 134–149 (2018)

[4] K.K. Pradhan, S. Chakraverty, Effects of different shear deformation theories on free vibration of functionally graded beams. Int. J. Mech. Sci. **82**, 149–160 (2014)

[5] J.N. Reddy, Nonlocal theories for bending, buckling and vibration of beams. Int. J. Eng. Sci. **45**, 288–307 (2007)

[6] B. Uzun, M.Ö. Yaylı, Nonlocal vibration analysis of Ti-6Al-4V/ZrO2 functionally graded nanobeam on elastic matrix. Arab. J. Geosci. **13**(4), 1–10 (2020)

Index

Printed in the United States
by Baker & Taylor Publisher Services